T0180153

Oral Biofilms and Modern Dental Materials

Andrei Cristian Ionescu
Sebastian Hahnel
Editors

Oral Biofilms and Modern Dental Materials

Advances Toward Bioactivity

Editors
Andrei Cristian Ionescu
Oral Microbiology and Biomaterials
Laboratory, Department
of Biomedical, Surgical, and Dental
Sciences
University of Milan
Milan
Italy

Sebastian Hahnel
Prosthetics and Dental Materials Clinic
Leipzig University
Leipzig
Germany

ISBN 978-3-030-67390-1 ISBN 978-3-030-67388-8 (eBook)
https://doi.org/10.1007/978-3-030-67388-8

This Springer imprint is published by the registered company Springer Nature Switzerland AG
The registered company address is: Gewerbestrasse 11, 6330 Cham, Switzerland

Introduction: Paradigm Shifts

This book is intended to outline the various aspects of the interactions between biofilms and modern dental materials and to highlight recent and potentially emerging paradigm shifts in the way we study and interpret these interactions. Biofilms are complex structures hosting microbial communities that support extraordinarily sophisticated interactions with the substrate they are colonising as well as with the outer microenvironment, and within the biofilm itself among the different microbial species inhabiting an ecological niche. These interactions aim to reach an equilibrium with the host and its microenvironment, ensuring the survival of the community over time in that particular ecological niche. Any factor that modifies this equilibrium in the sense of a detrimental effect to the host can be considered as dysbiosis. From this point of view, dental caries is an infectious disease originating from a disease of the biofilm itself, namely an imbalance between pathogenic and non-pathogenic species inhabiting the ecological niche. Considering this aspect, Chap. 1 summarises the essential characteristics of oral biofilms and processes involved in their development.

Regarding the aetiological processes involved in dental caries, all pathogenic microbial species involved feature acid production. The conventional wisdom is that local demineralisation of tooth structures by acids produced by cariogenic bacteria is the ultimate aetiological factor for caries formation. Demineralisation of hydroxyapatite initiates already at a pH of around 5.5, yet acidogenic biofilms reach values that are even lower than 4.0. Moreover, the accumulation of small, recurrent steps towards dysbiosis can also cause a drift of the equilibrium, as demonstrated by Philip Marsh [1]. With regard to this aspect, the relevance of saliva and the salivary pellicle cannot be overestimated. The salivary pellicle paves the ground for distinct microorganisms to firmly adhere to hard surfaces; thus, the salivary pellicle can impact the composition of a biofilm at least in its early colonisation phases. In this context, Chap. 2 summarises relevant properties of the salivary pellicle and its interaction with the surface of dental materials.

It might be amusing that the term "paradigm shift" was introduced by Thomas Kuhn, a physicist, philosopher, and historian of science, around the same time Bowen introduced the use of resin-based materials in dentistry. The term implies a fundamental change in the basic concepts and experimental practices of a scientific discipline, often driven not only by the introduction of new materials but also by a change in the protocols applied to a particular field. It regularly coincides with a complete change of perspective.

Perhaps the most used (and overused) example of paradigm shift is the introduction of Einstein's theory of relativity. That theory does not reject Newtonian physics, since, for extremely low speeds of light such as the ones that we are used to, the behaviour of the universe is just as well described by Newton's. Einstein, however, provided a completely new way to look at things, and tools for that, so far that it changed our perspective. In dentistry, the resin-based materials that were introduced by Bowen paved the ground for a paradigm shift towards adhesive and truly aesthetic dentistry, and towards the development of minimally invasive treatment concepts, as nicely summarised by Burke in his work from extension for prevention to prevention of extension [2].

In fact, resin-based materials are the materials of choice in modern minimally invasive dentistry. While relevant improvements in adhesive technology have relevantly paved the ground for the clinical success of these materials, computer-assisted design and manufacturing (CAD/CAM) techniques, as well as continuing advancements in 3D printing, will pave the ground for another paradigm shift as even complex rehabilitation can be planned and performed in an extensive, highly predictable and minimally invasive manner. Nevertheless, the main reasons for the failure, particularly of direct resin-based restorations, are secondary caries and material failure (fractures). While dramatic improvements have been made regarding the mechanical properties of the resin-based dental materials, secondary caries remains a relevant issue, particularly in patients with high caries risk. Until now, approaches to supplement resin-based materials with antibacterial properties have not led to the desired results. While for a long time the received opinion has been that secondary caries originates from gaps between the material and surrounding tooth tissues, we now know that this assumption is not correct unless the gap is exceedingly wide. Similarly to the primary disease, secondary caries results from an imbalance in the biofilm adherent to the surface of the restored tooth. As dental materials can now accurately reproduce some aspects of natural tooth tissues, the main aetiological factor in the onset of secondary caries might be the material itself. Concerning this aspect, current scientific theories regarding the aetiology of secondary caries and its implication with oral biofilms are discussed in Chap. 3.

The development of biofilms on the surface of dental materials and their individual properties can be studied using different experimental settings. The conventional wisdom is that in vivo/in situ studies feature the highest level of evidence and the results from in vitro studies can hardly be transferred into clinical settings, while studies performed using animal models are somewhere in between. Bioreactors, or artificial mouths, are sophisticated experimental systems which allow the analysis of biofilms by reproducing the conditions of the oral cavity in vitro. As these experimental systems may simulate many parameters of the oral environment, a reductionistic experimental approach can be employed as research methodology. Thus, the interactions between dental hard tissues, dental biomaterials, and oral biofilms can be studied extensively. The latest advancements in simulation techniques allow researchers to scale down the gap between laboratory models and clinical studies, including experimental models that are even closer to clinical settings than animal studies. In medicine, the possibility to accurately simulate clinical conditions is a relatively new way of looking at experimental models and is producing a paradigm shift in the way that experiments on dental materials are designed. Currently, several bioreactor models are regularly used to analyse biofilms on the surface of dental materials. These systems differ in design and complexity, and, unfortunately, there is no consensus on parameters such as the type of nutrient broth, flow speed, and the rate of shear stress. Thus, there is an urgent need for standardisation in order to make results from different groups comparable. With regard to this aspect, Chap. 4 provides insights into the latest advancements regarding the simulation of oral environments for analysing biofilm formation in vitro.

Biofilms on the surface of dental materials relevantly differ in terms of composition and structure from those adherent to natural tooth tissues. As explained in Chap. 5, the formation of biofilms on the surface of restorative materials is primarily impacted by surface properties, including surface roughness, nano- or microtopography, surface free energy, surface chemistry, and zeta potential. With regard to this aspect, surface properties do also influence the formation of the salivary pellicle and, thus, adhesion of pioneer microorganisms. Moreover, it has frequently been discussed that surface properties may influence microorganisms through several layers of the biofilm—mechanisms that we are only beginning to understand. For instance, signalling strategies inside the microbial community such as quorum sensing make microorganisms prepare for adhesion, including the expression of adhesins on their membrane. This process initiates already from a considerable distance to the substratum surface. Modern resin-based materials feature a complex composition and are tailored from a variety of ingredients with very distinct properties. Thus, adherence to the surface of these materials is more difficult to predict and interpret than to other dental materials. In this context, Chap. 6 summarises the current evidence on the interaction between resin-based materials and oral microorganisms in terms of biofilm formation.

Conventional dental materials are applied for permanent use in the oral cavity and have to face an extreme environment. In recent years, scientists discovered that they could degrade with time. This process includes both the adhesive interface, resulting from the activity of endogenous metalloproteinases, and the outer layers of the restoration, resulting from the activity of esterases delivered from the host and its biofilm. These processes may relevantly impact the mechanical and aesthetic properties of the material as well as its interface with natural tooth tissues. Chapter 7 provides a survey on the impact of the oral environment on the deterioration of dental materials and highlights current strategies for prevention of these processes.

In contrast to natural tooth tissues, the surface of dental materials is not capable of modulating the pH in acidic environments. Thus, biofilms on the surface of dental materials need to withstand low pH values for longer periods than biofilms on the surrounding natural tissues. This phenomenon fosters the prevalence of acid-tolerant species in the biofilm and underlines the fact that modulating the properties of a material can massively impact the composition and organisation of the biofilm adhering to its surface. As a result, current and future research focuses not only on the improvement of mechanical properties of dental materials but also on the optimisation of its interaction with host tissues and oral biofilms. From this point of view, it is our opinion and strong belief that a dental material will have to dynamically react to changing environmental conditions, being able to sustain a specific behavior and activity for a defined time. This comprehensive approach coincides with another paradigm shift in dental materials science.

As a consequence of the demineralisation process itself, natural tooth tissues may effectively buffer acidic conditions. Thus, dental caries might be regarded as an attempt of the organism to provide neutral pH conditions. While this is an entirely new perspective towards a disease that is still treated by attempting to eliminate biofilms, it appears that any treatment attempting to restore neutral pH values in close proximity to natural tooth tissues as quickly as possible after acidic challenge might relevantly affect the composition of the biofilm.

Bioactive materials have been available in many fields of dentistry for a long time, including materials for guided bone and tissue regeneration, dental implants, endodontic repair, and ion-releasing restorative materials such as glass-ionomer cements. In the surgical, periodontal, prosthetic, and endodontic field, clinical pressures on the advancement of such technologies have mainly been directed towards improved biocompatibility and healing as well as maximised predictability. Current strategies aim to produce dental materials with similar properties to natural tooth tissues, which can release ions and active compounds and can also be recharged. This approach allows healing of compromised natural tissues by the activity of the restorative biomaterial. Biomaterials with direct activity on biofilms, such as materials releasing active principles modulated by acidic conditions of the microenvironment in close proximity, may buffer and regulate the acidogenicity of a biofilm. While research has focused on the possibility to recharge these materials with active compounds for a long time with mixed results, recent studies have highlighted that biofilms may modulate the release and uptake of active compounds from a biomaterial surface by themselves. With regard to this aspect, the latest advancements in the development, chemistry, and performances of bioactive dental materials are summarised in Chaps. 8 and 9.

Moreover, antimicrobial dental materials that are being developed, currently being available to the clinicians, are based on two distinct strategies to influence biofilm formation. These strategies include either contact inhibition of microorganisms using immobilised bactericides or controlled release of antibacterial compounds. Chap. 10 reports on these approaches and their clinical effectiveness.

All these considerations underline that research on dental (bio)materials is exciting and evolving with extreme speed. However, we are convinced that the concepts and reasons beyond these evolutionary processes will produce changes in the next years that will influence the way we look at things for a long time.

We hope our readers will enjoy reading this book at least as much as our colleagues and we enjoyed writing it.

Sincerely,

Andrei Cristian Ionescu and Sebastian Hahnel

References

1. Marsh PD, Head DA, Devine DA. Prospects of oral disease control in the future—an opinion. J Oral Microbiol. 2014;6:26176.
2. Burke FJ. From extension for prevention to prevention of extension: (minimal intervention dentistry). Dent Update. 2003;30:492–8.

Contents

Oral Biofilms: What Are They?

1

Lakshman Samaranayake, Nihal Bandara,
and Siripen Pesee

Abstract

Biofilms are ubiquitous in nature. It is now known that within the oral ecosystem, bacteria and fungi mostly exist attached to surfaces, in the biofilm phase in contrast to their suspended or planktonic phase existence. Development of a biofilm from the initial seeding of organisms onto an oral substrate, such as enamel or a newly inserted appliance, to the climax community of a mature biofilm is a multiphasic process. A biofilm develops as soon as an appliance is introduced into the oral cavity, irrespective of the quality of the fabricated material or its manufacturing process. As plaque biofilms are the major etiologic agents of caries and periodontal disease, any dental biomaterial that suppresses the process would be superior and desirable than the traditional counterparts. This chapter outlines the various stages, and the factors that impact plaque biofilm development within an oral substrate or a surface of a biomaterial within the oral cavity. A basic understanding of this critical phenomenon would be valuable for fabricating "biofilm-retardant" new dental materials.

L. Samaranayake (✉)
Faculty of Dentistry, University of Hong Kong, Hong Kong, China (SAR)

Thamassat University, Bangkok, Thailand
e-mail: lakshman@hku.hk

N. Bandara
Oral Microbiology, University of Bristol, Bristol, UK
e-mail: nihal.bandara@bristol.ac.uk

S. Pesee
Oral Biology, Faculty of Dentistry, Thamassat University, Bangkok, Thailand

1.1 Introduction

The human body is composed of approximately hundred trillion cells, of which 90% comprise the resident microflora of the host and only 10% are mammalian eukaryotic cells. Bacteria are by far the predominant group of organisms in the oral cavity, and there are probably some 700 common oral species or phylotypes of which only 50–60% are cultivable in the laboratory. Of these, approximately 54% are officially named, 14% are unnamed (but cultivated), and 32% are known only as uncultivated phylotypes, or so-called uncultivable flora. Humans are not colonized at random and the microbial residents we harbor and provide shelter have coevolved with us over millions of years. This has led to the realization that the host and its residents together contribute to health and disease as a holobiont.

Although the coexistence and coevolution of microbes and humans have been known for decades it is only now we understand the massive complexity of the human microbiome due to the evolution of analytical tools and new genomic technologies. The latter, now miniaturized through microfluidics,

A. C. Ionescu, S. Hahnel (eds.), *Oral Biofilms and Modern Dental Materials*,
https://doi.org/10.1007/978-3-030-67388-8_1

dominated by next-generation sequencing (NGS) and third-generation sequencing (TGS) and the associated advances in bioinformatics have provided the scientific community with powerful tools to understand the role of the oral microbes in health and disease. This chapter provides a thumb sketch of the oral microbiome and biofilms, their development, functionality, and how they may interact in an oral environment replete with artificially introduced dental materials.

1.2 The Oral Microbiome

A perplexing array of organisms with diverse characteristics live in the oral cavity and the latter is arguably one of the most heavily colonized parts of the human body. This is due to its unique anatomical structures found nowhere else in the body, such as teeth and gingiva. Under normal circumstances this vast array of organisms including bacteria, archaea, fungi, mycoplasmas, protozoa, and a viral flora usually live in harmony in eubiosis with the host. But diseases such as caries and periodontal disease ensue when there is an ecological imbalance in the oral cavity due to either intrinsic or extrinsic causes, and a dysbiotic microbiome develops.

The totality of the oral microbes, their genetic information, and the oral environment in which they interact is called the oral microbiome whilst all living microbes constituting the oral microbiome are termed the oral microbiota. The oral microbiome in turn could be divided into three major sub-compartments as the oral bacteriome (bacterial component), oral mycobiome (fungal component), and the oral virome/virobiome (viral component) (Fig. 1.1). Apart from the above three distinct compartments of the oral microbiome, recent reports indicate the existence of a multitude of yet-to-be cultured (or uncultivable) ultrasmall bacteria that may fall within the bacteriome group, and these have been classified into a sub-sector called "*candidate phyla radiation* (CPR)" group. CPR group organisms are ultrasmall (nanometer range, compared to micron-scale bacteria) and highly abundant (>15% of bacteriome) group with reduced genomes and unusual ribosomes. They appear to be obligate symbionts, living attached to either the host bacteria or the fungi.

In recent studies of the structure, function, and diversity of human oral microbiome, evaluated using next-generation sequencing (NGS) technology mentioned above, it has been clearly shown that the oral microbiome is unique to each individual. It is indeed a "microbial thumbprint" of the individual. Even healthy individuals differ remarkably in the composition of the resident oral microbiota. Although much of this diversity remains unexplained, diet, environment, host

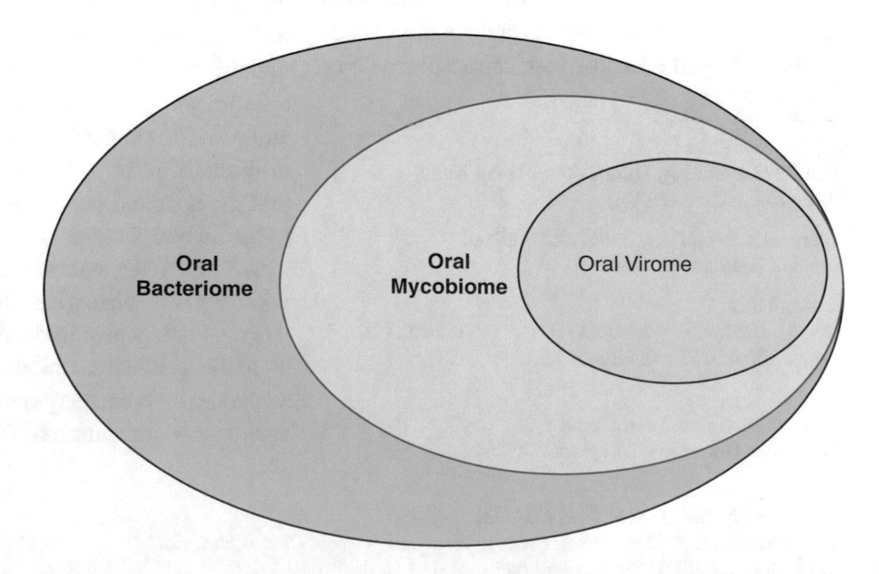

Fig. 1.1 The three major components of the oral microbiome: the relative size of the circles indicates the proportionality of the three components. (From Samaranayake L, *Essential Microbiology for Dentistry*, 5th Ed, Elsevier 2018, with permission)

genetics, and early microbial exposure have all been implicated in the constituent flora of the community that eventually stabilizes to form the so-called "climax community."

The oral microbiota exist either suspended in saliva as planktonic phase organisms or attached to oral surfaces, in the biofilm phase (also called the sessile phase) for instance as the plaque biofilm. In general, despite the high diversity in the salivary microbiome within and between individuals, little geographic variations can be noticed. Individuals from different parts of the world harbor similar salivary microbiota, indicating that host species is the primary determinant of the oral microbiome.

Some oral microbes are more closely associated with diseases than others (e.g., *Porphyromonas gingivalis*, a periodontal pathogen) although they commonly lurk within the normal oral flora without harming the oral health. This symbiosis between beneficial and pathogenic organisms is the key factor that contributes to the maintenance of oral health. IIn other words, when this homeostasis and the symbiotic equilibrium breaks down for example on taking broad-spectrum antibiotics, a state called dysbiosis sets in - leading to diseases such as caries, periodontal disease, and candidal infections. Essentially, in a dysbiotic microbiome the diversity and relative proportions of species or taxa within the microbiota are disturbed.

On occasions, specific microbes (mainly lactobacilli) could be administered to help restore a natural healthy microbiome in a given habitat (i.e., to convert a dysbiotic state to a symbiotic state). Such organisms are known as probiotics.

1.3 Biofilm Formation

Up to 65% of human infections are thought to be associated with microbial biofilms. Dental plaque on solid enamel surfaces is a classic example of a biofilm. As biofilms are ubiquitous in nature and form on hulls of ships, warm water pipes, dental unit water systems, and so on, their study has rapidly evolved during the past few decades, leading to many discoveries on communal behavior of microbes.

The sequence of events leading to oral biofilm formation is rather unique. When an organism first enters the oral cavity through either contaminated food particles or liquids, they are in the suspended planktonic phase. In order to survive in the hostile oral environment they need to adhere to either a biotic or an abiotic oral surface. Once the organisms adhere to the host surface they usually tend to aggregate and form intelligent communities of cells called biofilms. Biofilms are defined as a complex, functional community of one or more species of microbes, encased in an extracellular polysaccharide matrix and attached to one another or to a solid surface (such as a denture prosthesis, enamel, filling materials).

Biofilms are intelligent, functional communities. Structurally, they are not flat and compressed, but comprise a complex architecture with towers and mushroom - or dome-shaped structures with water channels that permit transport of metabolites and nutrients. Bacteria in biofilms maintain the population composition by constantly secreting low levels of chemicals called quorum-sensing molecules (e.g., homoserine lactone, autoinducers), which tend to repulse incoming bacteria or activate the communal bacteria to seek new abodes. Further, specific gene activation may lead to production of virulence factors or reduction in metabolic activity (especially those living deep within the matrix).

It is now known that infections associated with biofilms are difficult to eradicate as sessile organisms in biofilms exhibit higher resistance to antimicrobials than their free-living or planktonic counterparts. The reasons for this phenomemon appear to be (1) protection offered by the extracellular polysaccharide matrix from the host immune mechanisms, (2) poor penetration of the antimicrobials into the deeper layers of the biofilm, (3) degradation of the antimicrobials as they penetrate the biofilm, (4) difference in pH and redox potential (E_h) gradients that is not conducive for the optimal activity of the drug, and (5) gene expression leading to more virulent or resistant organisms. Such bacteria which are encased within the biofilm matrix and express resistance to antibiotics or antiseptics are called "persisters."

1.4 Plaque Biofilm

Plaque biofilm is found on dental surfaces and appliances as well as on dental restorations of various kinds (e.g., amalgam, composites, ceramic). Biofilm growth is particularly common in the absence of oral hygiene. In general, it is found in anatomic areas protected from the host defenses and mechanical removal, such as occlusal fissures and edges of poorly prepared dental restorations. In the latter instance, biofilm plaque growth is associated with secondary caries, and depending on the antomic location of the restoration, gingivitis, and even periodontitis. Peri-implantitis and peri-implant mucositis are primarily due to plaque biofilms that develop on implants which are improperly maintained with poor oral hygiene.

1.5 Composition

The organisms in dental biofilm are surrounded by an organic matrix, which comprises about 30% of the total volume. The biofilm matrix is derived from the products of both the host and the biofilm constituents, and it acts as a food reserve and as a cement, binding organisms both to each other and to various surfaces.

The microbial composition of dental plaque biofilm can vary widely between individuals: some people are rapid plaque formers; others are slow. Even within the same individual, there are also large variations in plaque composition. These variations may occur at different sites on the same tooth, at the same site on different teeth, or at different times on the same tooth site.

1.6 Formation

The dental biofilm formation is a complex process comprising several stages. The first step is the salivary pellicle formation, where adsorption of host and bacterial molecules to the tooth surface forms an acquired salivary layer. The pellicle formation starts within hours after cleaning the tooth, and it is composed of salivary glycoproteins, phosphoproteins, lipids, and components of gingival crevice fluid. Remnants of cell walls of dead bacteria and other microbial products (e.g., glucosyltransferases and glucans) are also part of the pellicle. This stage is essential for the plaque biofilm formation, because the oral bacteria do not directly colonize the mineralized tooth surface; they initially attach to this pellicle. Salivary molecules alter their conformation after binding to the tooth surface, exposing receptors for bacterial attachment. These receptors interact with bacterial surface adhesins in a very specific manner that explains the selective adherence of bacteria to enamel.

The transport of bacteria to the vicinity of the tooth surface before attachment occurs by means of either natural salivary flow, Brownian motion, or chemotaxis. The proximity to mineralized surfaces allows the attachment of earilest arriving bacteria called pioneer colonisers (0–24 h). This important group of bacteria comprising the initial or pioneer colonizers are gram-positive cocci and rods, mainly streptococcal species (*S. sanguinis*, *S. oralis*, and *S. mitis*), and to a lesser extent, *Actinomyces* spp. and gram-negative bacteria (e.g., *Haemophilus* spp. and *Neisseria* spp.).

The interactions between the microbial cell surface, the initial or pioneer colonizers, and the pellicle-coated tooth involve weak physicochemical forces (van der Waals forces and electrostatic repulsion), which represents a reversible phase of net adhesion. These interactions may rapidly become strong stereochemical reactions between adhesions on the microbial cell surface and receptors on the acquired pellicle. This phase is an irreversible phase in which polymer bridging between organisms and the surface helps to anchor the organism, after which they multiply on the virgin surface. Doubling times of plaque bacteria can vary considerably (from minutes to hours), both between different bacterial species and between members of the same species, depending on the environmental conditions.

After the establishment of pioneer colonizers, the next stage involves the co-adhesion and growth of attached bacteria, leading to the formation of distinct microcolonies (4–24 h). During this phase, the biofilm is not uniform in thickness, varying from sparsely colonized to almost full surface coverage. The biofilm grows basically by cell division, with the development of columnar microcolonies perpendicular to the tooth surface. Within one day, the tooth surface is almost completely covered by a blanket of microorganisms.

Some constituents of the dental plaque produce components of the biofilm matrix, such as polysaccharides. The biofilm matrix, in turn, not only gives support for the structure of the biofilm but is also biologically involved in retaining nutrients, water, and enzymes within the biofilm.

Between 24 and 48 h, the biofilm becomes thicker. Adhesins expressed by the secondary colonizers bind to receptors on the cell surface of pioneer microbes, producing a co-aggregation or co-adhesion of microorganisms. This continuous adsorption of planktonic microbes from saliva, in addition to cell division, contributes to the expansion of the biofilm. In the surface layer, co-aggregation of different species creates "corncob" structures.

As the dental biofilm develops, the metabolism of the initial colonizers modifies the environment in the developing biofilm, creating local conditions that are either more attractive to later (secondary) colonizers or increasingly unfavorable to the pioneer group, for example, by making it more anaerobic after their consumption of oxygen or accumulating inhibitory metabolic products.

All these environmental changes lead to a gradual replacement of the initial or pioneer colonizers by other bacteria more suited to the modified habitat; this process is termed microbial succession (1–7 days). This sequence of events increases species diversity in the dental plaque, concomitant with continued growth of microcolonies. A progressive shift is observed from mainly aerobic and facultatively anaerobic species (mainly streptococci) in the early stages of biofilm formation to a situation with predominance of facultatively and obligatory anaerobic organisms, gram-negative cocci and rods, fusobacteria, spirochetes, and actinobacteria (especially *Actinomyces* species) after 9 days (Fig. 1.2).

A mature biofilm can be found after one week or more. The plaque mass reaches a critical size at which a balance between the deposition and loss of plaque bacteria is established, characterizing a climax community. In an old biofilm, structural changes can be seen at the bottom of the dental plaque, for example. The outer part of a mature biofilm is usually loosely structured with various compositions, from sphere-shape distribution of

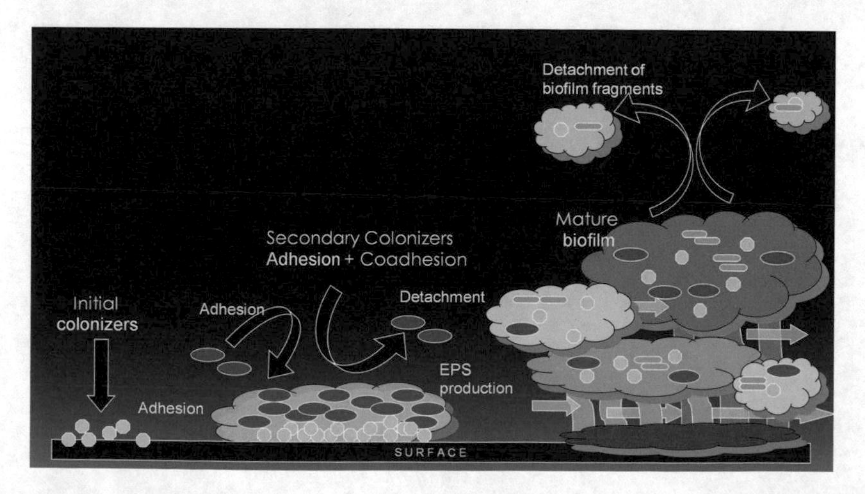

Fig. 1.2 A schematic diagram depicting the various developmental stages of a biofilm from the initial adherent phase (left) of the organisms to gradual maturation and subsequent fully developed polymicrobial biofilm (extreme right); green arrows = water channels. (From Samaranayake L, Essential Microbiology for Dentistry, 5th Ed, Elsevier 2018, with permission)

one type of organism to multispecies outer microflora with parallel distribution. The bacteria that colonize this climax community may detach and enter the planktonic phase (i.e., suspended in saliva) and may be transported to new colonization sites, thus restarting the whole cycle.

The colonization of root surfaces follows similar principles to those outlined above for enamel surfaces. However, the development of plaque on root surfaces occurs more rapidly due to the uneven surface topography. Despite this difference, regardless of the type of tooth surface, enamel and cementum share the same initial colonizers.

1.7 Biofilm Functionality

The oral biofilms function as a microbial community and collectively display properties that favor their formation and persistence in the oral cavity. Many of these properties make the microorganisms within a biofilm more resistant to drugs (antibiotics, antifungals) in comparison to their planktonic counterparts. In the biofilm, microbial cells interact both with each other via cell signaling systems and with other species through conventional synergistic and antagonistic biochemical interactions. Cell-cell communication and coordinated population-based behavior among members of a biofilm involve quorum sensing, a system of signaling molecules (e.g., homoserine lactone, autoinducers that increase according to the cell density). This ability to send, receive, and process information allows organisms within microbial communities to act as multicellular entities and increases their chances of survival in complex environments. Furthermore, for the entry of nutrients and the efflux of metabolites the mature biofilm appears to possess an elegant transport system comprising so-called water channels (Fig. 1.3), although some argue that the channels are not particularly suited for this purpose.

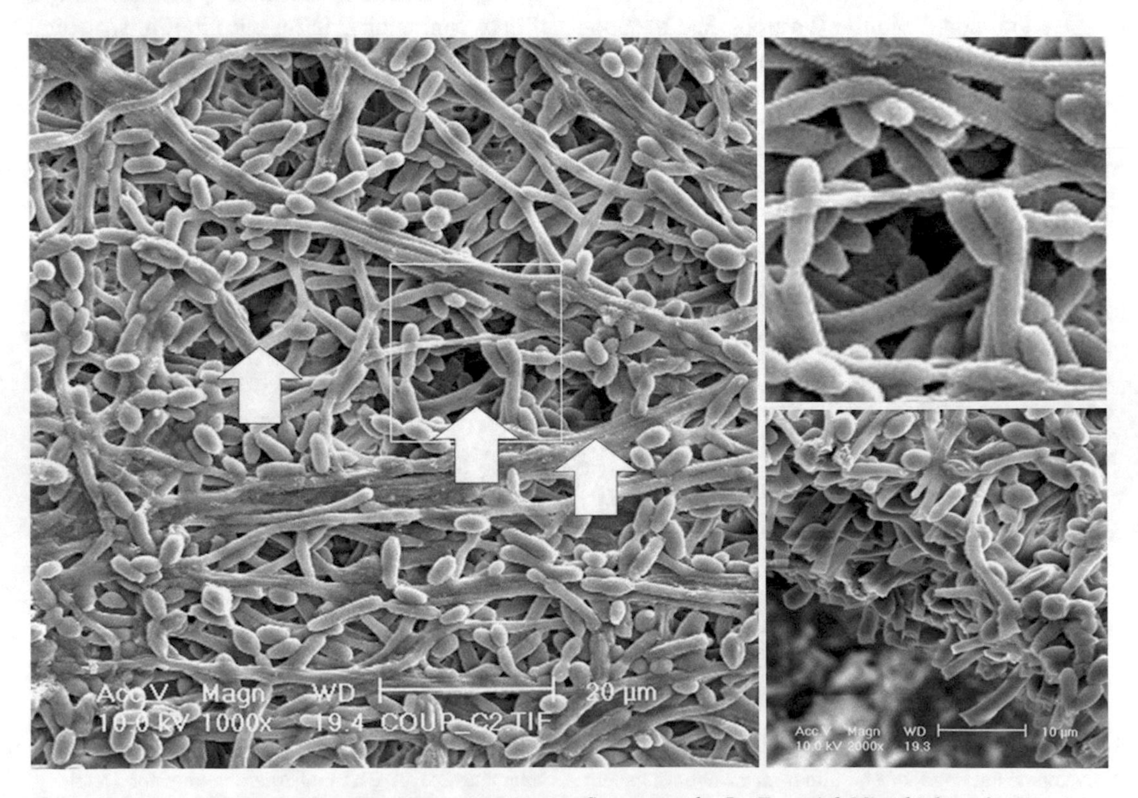

Fig. 1.3 A mature biofilm of *Candida albicans* showing water channels (white arrows) that mediate metabolite and nutrition transfer to and from the biofilm. (From Samaranayake L, *Essential Microbiology for Dentistry*, 5th Ed, Elsevier 2018, with permission)

Acknowledgement Professor Samaranayake gratefully acknowledges the Thammasat University of Thailand for the award of a Bualuang ASEAN Chair Professorship to support this research.

Further Reading

Kilian M, Chapple I, Hannig M, et al. The oral microbiome—an update for oral healthcare professionals. Br Dent J. 2016;221:657–66. https://doi.org/10.1038/sj.bdj.2016.865.

Lang NP, Mombelli A, Attstrom R. Ch. 3: Dental plaque and calculus. In: Clinical periodontology and implant dentistry. 3rd ed. Copenhagen: Munksgaard; 1997.

Listgarten MA. The structure of dental plaque. Periodontology 2000. 1994;5:52–65.

Marsh PD, Martin MV. Oral microbiology. 5th ed. London: Butterworth-Heinemann; 2009.

Samaranayake L. Essential microbiology for dentistry. 5th ed. Edinburgh: Elsevier; 2018.

Samaranayake L, Matsubara VH. Normal oral flora and the oral ecosystem. Dent Clin. 2017;61(2):199–21.

2 The Importance of the Salivary Pellicle

Matthias Hannig

Abstract

The acquired salivary pellicle forms rapidly on any type of surface exposed to the oral environment. The adsorption of salivary biomolecules onto solid surfaces is a very complex process determined by the interplay of the substrate, proteins, and other components from saliva as well as the surrounding environmental conditions, e.g., ionic strength, shear forces, and nutrition. The acquired pellicle will act as a multifunctional protective layer for the underlying surface providing lubrication and partial protection against wear. In contrast to the huge number of publications covering the pellicle's formation, composition, and functional properties on the tooth surface, only very sparse information is available on the salivary pellicle formed in situ or in vivo on dental materials. The pellicle layer will contribute to some extent to the passivation of the surface chemistry and masking of the surface texture of dental materials. Currently, the proteome of the acquired pellicle layer formed in situ on ceramics has been characterized in detail, thereby identifying more than 1180 proteins as well as 68 main proteins contributing to the formation of the 3-min pellicle. Open questions and challenges of the salivary pellicle's impact in restorative dentistry and material's science are highlighted. Future directions in research considering pellicle formation, composition, and function on dental materials are discussed.

M. Hannig (✉)
Clinic of Operative Dentistry, Periodontology and Preventive Dentistry, Saarland University, Homburg, Germany
e-mail: matthias.hannig@uks.eu

2.1 Introduction

Intensive research has been conducted over the last six decades to gain a detailed understanding of the acquired enamel pellicle layer, i.e., the salivary pellicle formed on the natural tooth enamel surface (see reviews [1, 2]). Numerous in vitro as well as in situ and in vivo studies indicate that the process of pellicle formation onto the enamel surface is related to the selective adsorption of salivary biopolymers (see reviews [1, 2]). Thus, it is very likely that also on different dental restorative materials the acquired pellicle formation will take place in a highly selective manner resulting in the formation of material-specific conditioning films. Although this consideration appears quite reasonable, the scientific literature is very controversial regarding selective and specific pellicle formation on different dental materials [3]. Mainly based on the results of in vitro studies, it has been shown that the salivary proteinaceous layer adsorbed onto different materials will vary in composition, structure, and function in a surface-dependent manner [4–8].

A. C. Ionescu, S. Hahnel (eds.), *Oral Biofilms and Modern Dental Materials*,
https://doi.org/10.1007/978-3-030-67388-8_2

This chapter highlights the present knowledge on the formation, composition, structure, physicochemical properties, and function of the salivary pellicle adsorbed on enamel and dental materials under the influence of the oral environment. Thereby, this review aims to give an overview regarding the impact of dental materials on salivary pellicle formation and function, and vice versa the influence of the salivary pellicle on the properties of biomaterials under oral conditions with emphasis on current research and future scientific perspectives.

2.2 Pellicle Formation

Under in vivo conditions the acquired pellicle layer is formed nearly immediately on any biomaterial's surface getting in contact with the oral fluids [9–11]. Components originating from saliva, gingival crevicular fluid, mucosal transudate, bacteria, and nutrition will contribute to the formation of the pellicle layer [2]. Salivary proteins generally interact with the solid substrate via a complex combination of various simultaneously acting intermolecular forces [3]. Under oral conditions, spontaneous adsorption of salivary proteins will take place in contact with any solid surface due to the heterogeneous molecular properties of the proteins that include hydrophobic and hydrophilic groups, positive and negative charges, dipoles, and hydrogen-binding groups. Because of its specificity and complex molecular structure, it had been assumed that each protein might exhibit its own "personality" during the adsorption process [12]. Which proteins are preferentially and predominantly adsorbed onto the solid substrate, their conformations, and binding forces are decisive for the properties and function of the pellicle layer. Regarding the time course, pellicle formation on solid surfaces exposed intraorally involves a fast early stage lasting for minutes followed by a slow continuous stage spanning hours [13].

Dental materials exposed intraorally might affect the formation and composition of the adsorbed salivary pellicle layer in many ways by their chemical composition, surface energy, surface roughness, and topography. However, analysis of in situ pellicle formation on PMMA, amalgam, and gold specimens by ex vivo contact angle measurements revealed that the wettability of all materials increased due to the adsorbed pellicle layer [14]. Within 5–20 min of in situ pellicle formation time the effect of the original surface activity of the materials was effectively sealed off [14]. Experiments conducted with pellicle layers formed in situ on model materials with highly differing wettability (mica, silicon wafer, graphite) indicate that even within 10–30 s of intraoral exposure the wettability of the pellicle-covered surfaces will reach a similar level [11]. Also recent in vitro experiments demonstrate that formation of the salivary pellicle contributes to the equalization of surface charges and surface free energy comparing acrylic and titanium surfaces [15]. It can be concluded that pellicle formation will level off the original surface free energy of different materials yielding a homogenizing effect and facilitating a physiological interface between restorations and the oral environment [3, 14, 16].

Pellicles which were formed for 1 h in vitro on substrates of different wettability reveal no major differences in their composition [4]; thus, it has been assumed that variations in substratum wettability will have a major impact on the organization and conformation of the adsorbed pellicle components rather than on the pellicle's composition [4]. This very interesting assumption needs further validation by studies involving in vivo-formed pellicle layers on dental materials.

2.3 Composition of the Pellicle

The acquired pellicle constitutes of proteins, peptides, lipids, carbohydrates, and dietary macromolecules adsorbed from the oral fluids [1, 2]. In recent years, the protein composition of the acquired pellicle has been within the focus of several in vivo and in situ studies, applying modern label-free liquid chromatography-electrospray ionization tandem mass spectrometry for protein identification and characterization of the acquired pellicle's proteome [17–21]. These studies revealed a very diverse and complex composition of the pellicle formed on the natural tooth surface [17, 20, 21]. More than 360 proteins were identified contributing to the formation of the 2-h pellicle layer on tooth enamel in vivo [21]. Important topic differences in the proteomic profile of the 2-h acquired pellicle were recorded according to the pellicle's intraoral region of formation [21]. These differences in the proteome of the acquired pellicle across various sites of the dentition are caused by variations in the local availability of salivary proteins [2]. Furthermore, diet and nutrition will contribute to the pellicle's composition [17], at least influencing temporarily the protein profile of the pellicle.

Using a data mining analysis it has been demonstrated recently that there are specific differences between enamel pellicle and salivary proteins [22], which corroborates the experimental findings that the pellicle is formed by selective adsorption of salivary proteins onto the tooth surface. The differences between salivary proteins and pellicle proteins were shown to be related to the proteins' molecular size distribution, with pellicle proteins being significantly smaller with average molecular weights of mainly 10–20 kDa [22]. This finding based on bioinformatics analyses is in line with the well-accepted concept that pellicle formation on the enamel surface is based on selective adsorption of salivary proteins. Furthermore, a high cross-linking potential has been revealed for pellicle proteins [22], which might be relevant to get the pellicle's physiological structural organization and functional properties. Nevertheless, the proteomes of saliva and pellicle are rather similar in many aspects [22]. This result of the data mining analyses is not surprising as the pellicle's proteome (mainly) constitutes a subgroup of the salivary proteome [22].

The vast majority of the data related to the proteome of the in vivo- or in situ-formed acquired pellicle layer is based on pellicle samples pooled from several volunteers, neglecting the fact that the individually formed pellicle might reveal also an individual proteomic profile. Two recently published studies intended to characterize the individual pellicle's proteome instead of using pooled pellicle samples from up to 24 volunteers [18, 23]. In both studies the acquired pellicles were formed in situ in the oral cavity for 3 min onto ceramic surfaces. Ceramic was chosen as the substrate for pellicle formation in order to guarantee well-defined surfaces without biological variations for the adsorption of the pellicle layer. The results of these studies indicate that the proteomic profile of the 3-min in situ-formed pellicle layer is characterized by a much higher diversity of adsorbed proteins than ever reported before, and reveals a distinct individual-dependent protein composition. Overall, a total of 1188 different proteins and peptides were identified in the 3-min pellicles formed in situ on ceramic specimens [23]. Interestingly, 68 proteins were detected to be present in the proteomic profiles of all 24 individuals. These proteins were also identified in a previous study with four volunteers [18], and apparently represent the base proteins of the acquired salivary pellicle [23]. Among these 68 proteins identified in the 3-min in situ pellicle on ceramics are serum albumin, alpha-amylase, carbonic anhydrase 6, cystatins, lactoperoxidase, lactoferrin, lysozyme, mucin 7, proteins S100-A8 and A9, proline-rich proteins, or transglutaminase, which had been identified as constituents of the acquired enamel pellicle in previously published studies (see reviews [1, 2, 17, 20, 21]). Comparison of the individual pellicle proteomes besides the 68 common proteins revealed high inter-individual differences indicating that the proteomic profile of the individual pellicle might depict an individual fingerprint-like protein composition [23]. Future proteomic analyses of in situ-formed pellicle layers are necessary to elucidate the impact of longer formation times on the proteomic profile of the individual pellicle, and to clarify potential substrate-dependent influences of dental materials on the proteome of the salivary pellicle.

2.4 Ultrastructure of the Pellicle

The salivary pellicle formed on enamel has been characterized by transmission electron microscopic (TEM) investigations as a multilayer structure consisting of an inner compact, tightly bound basal layer, and an outer less dense layer of scattered globular agglomerates [24, 25]. The surface microstructure of the pellicle was described by high-resolution scanning electron microscopy as well as by atomic force microscopy as a spongy meshwork of protein aggregates [1, 11].

Up to now only few studies have been published regarding high-resolution imaging and visualization of the pellicle layer formed in situ on solid surfaces (dental materials) beside enamel [11, 24]. TEM investigations performed more than 20 years ago indicated that the principal ultrastructure and the time-dependent morphogenesis of the salivary pellicle formed in situ on a broad range of dental materials did not reveal distinct differences [24]. These experiments were performed under well-defined conditions, considering the preparation of the samples from various dental materials, their surface polishing procedure, and their hydration state before 2- or 6-h intraoral exposure in the oral cavities of three volunteers using removable acrylic splints [24]. The overall conclusion from these TEM analyses is that salivary pellicles are formed with high ultrastructural similarity on all solid surfaces exposed to the oral environment.

2.5 Function of the Pellicle

The pellicle plays an important role in the maintenance of oral health. As a physiological mediator the pellicle layer will determine and modify all interfacial interactions taking place at the tooth-saliva interface [1, 2]. The pellicle acts as a regulator of dental hard-tissue mineral homeostasis, and provides hydration, lubrication, and protection of the underlying substratum against chemical and mechanical wear and degradation at the tooth-saliva interface [2]. The physiological impact of the pellicle will be dependent on its composition as well as the ultrastructural orientation, cross-linking, and binding forces of the adsorbed biomacromolecules [2].

Under oral conditions, tooth surfaces are constantly exposed to chemical and mechanical degradation processes which are attenuated and regulated by the omnipresent salivary pellicle. Although the modulating effect of the salivary pellicle on all interfacial interactions taking place on restoration surfaces is of high clinical relevance, very few results were published, yet, covering this topic with special emphasis on dental materials [26, 27]. These sparse findings indicate that the salivary pellicle may protect the surface of restorative materials against erosive wear [27].

2.6 Pellicle and Corrosion of Metallic Dental Materials

Corrosion resistance of a dental alloy plays a key role considering its biocompatibility under the influence of the oral environment. However, only very limited scientific information is available regarding the impact of the natural pellicle on corrosion processes at the surface of metallic restorations or titanium implants, although it seems reasonable that the omnipresent acquired pellicle will act as a passivating layer to the underlying surface under oral conditions.

The corrosion behavior of numerous metallic dental materials had been studied mainly using different types of artificial saliva composed of ion solutions without any added proteins or mucins. However, in vitro studies have clearly demonstrated that protein solutions will inhibit or reduce the corrosion of alloys [28–30]. Based on the results with model proteins like bovine serum albumin (BSA), protein adsorption has been reported to enhance the corrosion resistance of implant materials [31, 32]. BSA coatings prevent the selective dissolution of the vanadium-rich ß-phase of Ti_6Al_4V alloy, thereby increasing the corrosion performance of the Ti_6Al_4V surfaces [32].

Only very rarely natural saliva was added to mimic the oral situation for corrosive degradation of dental materials under more realistic conditions [33–35]. Already in 1955 [35] an inhibiting effect of the natural saliva on the in vitro corrosion of several different metals had been reported. The corrosion profile of amalgam exposed in whole saliva is decidedly different from the corrosion behavior of amalgam polarized in an artificial saliva ionic solution under in vitro conditions [33]. These findings relativize and limit the value of investigations on corrosion of dental metallic materials in electrolytes other than natural saliva, and exclude the extrapolation of results obtained in nonsalivary media to the in vivo situation [34]. Only corrosion experiments performed under most realistic conditions will be able to produce reliable results regarding the in vivo corrosion behavior of metal alloy dental materials [34]. Håkansson et al. [34] could provide evidence by their ex vivo studies that under "realistic conditions" very low corrosion currents can be expected for amalgams in natural saliva, and the formation of "barrier films" on dental materials in natural saliva plays a decisive role for the stability of the materials under in vivo conditions.

To the best knowledge of the author there has been only one study published, in which corrosion experiments were conducted ex vivo with in vivo-formed salivary pellicle layers on metallic dental materials [36]. These experiments indicate that the 1-h in situ pellicle formation does not influence the galvanic corrosion behavior of amalgam induced by contact with a casting gold alloy. However, the 1-h pellicle layer formed on the gold alloy substantially reduced the corrosion of the amalgam, clearly underlining the relevance of the acquired pellicle in corrosion processes.

Based on the published results, it can be summarized that the detailed mechanism by which the adsorbed protein layer affects the corrosion behavior of dental alloys still lacks appropriate analysis at the micro, nano, and molecular level, thereby considering carefully the importance and relevance of the in vivo-formed acquired pellicle layer.

2.7 Pellicle and Lubrication

Intraoral lubrication is important to maintain physiological functions as deglutition and mastication as well as to facilitate and support speech [37]. The lubricating properties of the salivary pellicle will mainly depend on the tenacity and mechanical properties of the inner basal pellicle layer [38–40]. By means of colloidal probe atomic force microscopy it has been demonstrated that the presence of an in vitro-formed salivary pellicle layer reduces the friction coefficient between silica surfaces by a factor of 20 [37]. Thereby, the lubricating effect might be explained by full separation of the sliding surfaces due to the adsorbed pellicle layer. Pellicle components related to lubrication include mucins, statherin, and proline-rich proteins [41]. In particular, acidic proline-rich protein 1 might be of major relevance for the lubricating capacity of the pellicle [41]. However, only few in situ-gained data are available pointing to the physiological function of the pellicle as a lubricant [38]. Evidence had been provided that the in situ-formed salivary pellicle in fact decreases abrasive wear of enamel and dentin during daily toothbrushing [38].

In recent years, attempts have been undertaken in order to characterize the salivary pellicle's rigidity, tenacity, and viscoelastic properties. The friction coefficient and related wear loss of enamel have been shown to decrease significantly due to the formation of the salivary pellicle under in vitro conditions [39]. Thus, already the initially formed 1-min pellicle layer will exhibit an excellent lubricating effect on the enamel surface [39]. In addition, in vitro experiments with salivary pellicle layers formed on enamel over 60 min reveal that the shear energy between salivary pellicle and enamel increased exponentially with increasing adsorption time [40]. The adhesion force between the initial salivary pellicle and the enamel surface is more than twice higher than that between the initially formed pellicle and the outer pellicle layer [40].

Based on nanoindentation measurements performed on the 1-min in vitro-formed enamel salivary pellicle layer, the pellicle's intrinsic nanohardness was calculated with about 0.52 GPa, while the nanohardness of the underlying enamel was calculated as about 4.88 GPa [42]. These findings clearly indicate that the pellicle layer might act as a viscoelastic layer on the tooth surface. Furthermore, the viscoelastic properties of the in vitro-formed 60-min pellicle were shown to depend on the substrate with lowest values measured on hydroxyapatite compared to other substrates as titanium, gold, zirconia, or silica [5]. However, data regarding the lubricating effect and potential wear protection by the salivary pellicle formed on restorative or prosthetic dental materials under in vivo conditions are lacking, yet.

Moreover, the lubricating properties of the salivary pellicle might be of clinical relevance considering the complex abutment-screw-implant interfaces. According to the experiments performed by Bordin et al. [43] saliva and related pellicle formation might jeopardize the biomechanical behavior of dental implant-supported prosthodontic restorations. The presence of saliva at the complex abutment-screw-implant interfaces will cause a shift in the friction coefficient between the surfaces, thus contributing to decreased stress and insufficient tensile force in the screw [43]. An experimental 4-h salivary pellicle layer was demonstrated by tribometrical measurements to decrease the friction coefficient at the titanium-titanium interface, while increasing the friction coefficient between titanium and zirconia [43]. The reduced friction of the screw might contribute to an increased micromotion at the implant-abutment interface [43]. The result could be loosening of the abutment screw which is one of the most common problems in single-implant rehabilitations.

2.8 Pellicle Formation and Surface Roughness

Nano-sized porosities, scratches, gaps, fractures on the surface of the tooth, or restorative materials will be filled and masked by the adsorption of salivary biopolymers under oral conditions [25]. It has been shown in vitro that roughening of titanium surfaces will result in higher interactions with salivary proteins as compared to smooth surfaces [44]. Salivary conditioning films will significantly reduce the roughness of resin composites, thereby reducing the adhesion forces of streptococci under in vitro conditions [45]. Nevertheless, also in the presence of the in vitro-adsorbed salivary film, rougher composite surfaces exert stronger streptococcal adhesion forces than smooth surfaces indicating that the pellicle layer could only partially mask the surface roughness [45]. In accordance with these in vitro results, it has been shown that formation of the salivary pellicle under oral conditions will not completely compensate the influence of the surface topography of non-polished implant materials on bacterial adherence during 2 h of in situ biofilm formation [46].

Interestingly, the surface topography (size and depth of surface irregularities or porosities) might play a more important role for the extent of in vitro biofilm formation in the presence of saliva than the measured surface roughness which quantitatively characterizes the mean roughness of the surface [47]. Recent in vitro experiments indicate that interfacial curing conditions and related differences in surface topography and chemistry will significantly influence biofilm formation on resin-based composites [48]. However, these effects will be counterbalanced up to nullified by saliva preconditioning of the resin composite surfaces underlining the highly relevant impact of the pellicle layer [48].

2.9 Pellicle and Antimicrobial Properties of Dental Materials

Prevention of bacterial adherence and killing of bacteria, either by coatings that release antibacterial substances or by surface-associated (contact) mechanisms, are the most prevalent antimicrobial and anti-biofilm approaches [49]. However, all surface-related strategies might suffer from reduced efficacy due to the adsorption of biopolymers from body fluids [26, 49]. Antimicrobial coatings in dental applications may be compromised under oral conditions by salivary protein adsorption which influences the kinetics of active agent release and their efficacy [50–52]. For example, the antibacterial activity of resin composite filler particles containing the bactericide MDPB is attenuated by saliva pretreatment [50], and salivary pellicles decrease the antibacterial efficacy of DMADDM and nanosilver-containing antibacterial dental adhesives [51]. In addition, the bactericidal activity of pyridinium group-containing methacrylate monomers immobilized on silicon wafers was found to be significantly reduced due to the adsorption of salivary proteins [52].

In this context, it is also really surprising that the effect of the salivary pellicle on fluoride release from glass ionomer cements has not been studied in detail, yet. One in vitro study indicates that exposure to saliva for only 10 min diminishes the fluoride release from glass ionomer cement specimens [53]; however, studies focusing on the pellicle's effect on fluoride release under oral conditions are completely lacking.

2.10 Conclusions and Outlook

Based on the present knowledge the following conclusions might be justified considering the process of initial bioadhesion on dental materials under oral conditions:

- All surfaces exposed to the oral environment are readily covered by the salivary pellicle within a few minutes. Pellicle layers are regularly present on commonly used dental materials.
- Formation, composition, binding forces, function, stability, and tenacity of the pellicle will be influenced on the one hand by the physicochemical properties of the solid substrate, and on the other hand by the ambient oral environment, its composition, and local availability of macromolecules.
- The pellicle acts as a mediator to establish a physiological interface on solid surfaces exposed in the mouth. Thereby, physicochemical differences between different dental restorative materials are levelled out and equalized to a certain level.
- Acquired salivary pellicle formation will influence the biological behavior of dental materials under oral conditions, and thus might improve the biocompatibility of orally exposed materials.
- Testing and understanding of dental biomaterials necessarily need to consider the existence of the omnipresent, ubiquitous pellicle when designing the study.
- Future studies should be directed on the elucidation of the composition and function of the salivary pellicle formed on dental materials under in vivo conditions.
- It will be of utmost scientific interest and significance to measure and characterize the mechanical properties as well as the adhesion forces of the pellicle layers formed on different dental materials.
- These data will provide new insights into the process of pellicle formation, and would also be of importance for the development of new anti-biofilm strategies based on pellicle modifications.

Acknowledgement This work has been supported by the German Research Foundation (DFG, SFB 1027).

References

1. Hannig M, Joiner A. The structure, function and properties of the acquired pellicle. Monogr Oral Sci. 2006;19:29–64.
2. Siqueira WL, Custodio W, McDonald EE. New insights in the composition and functions of the acquired enamel pellicle. J Dent Res. 2012;91:1110–8.
3. Hannig C, Hannig M. The oral cavity: a key system to understand substratum-dependent bioadhesion on solid surfaces in man. Clin Oral Investig. 2009;13:123–39.
4. Aroonsang W, Sotres J, El-Schick Z, Arnebrant T, Lindth L. Influence of substratum hydrophobicity on salivary pellicles: organization or composition? Biofouling. 2014;30:1123–32.
5. Barrantes A, Arnebrant T, Lindh L. Characteristics of saliva films adsorbed onto different dental materials studied by QCM-D. Colloids Surf. 2014;442:56–62.
6. Santos O, Lindh L, Halthur T, Arnebrant T. Adsorption from saliva to silica and hydroxyapatite surfaces and elution of salivary films by SDS and delmopinol. Biofouling. 2010;26:697–710.
7. Svendsen IE, Lindh L. The composition of enamel salivary films is different from the ones formed on dental materials. Biofouling. 2009;25:255–61.
8. Weber F, Barrantes A. Real-time formation of salivary films onto polymeric materials for dental applications: differences between unstimulated and stimulated saliva. Colloids Surf B Biointerfaces. 2017;154:203–9.
9. Baier RE, Glantz PO. Characterization of oral in vivo films formed on different types of solid surfaces. Acta Odontol Scand. 1978;36:289–301.
10. Ericson R, Pruitt K, Arwin H, Lundström I. Ellipsometric studies of film formation on tooth enamel and hydrophilic silicon surfaces. Acta Odontol Scand. 1982;40:197–201.

11. Hannig M, Döbert A, Stigler R, Müller U, Prokhorova SA. Initial salivary pellicle formation on solid substrates studied by AFM. J Nanosci Nanotechnol. 2004;4:532–8.
12. Norde W. My voyage of discovery to proteins in flatland … and beyond. Colloids Surf B. 2008;61:1–9.
13. Güth-Thiel S, Kraus-Kuleszka I, Mantz H, Hoth-Hannig W, Hähl H, Dudek J, Jacobs K, Hannig M. Comprehensive measurements of salivary pellicle thickness formed at different intraoral sites on Si wafers and bovine enamel. Colloids Surf B. 2019;174:246–51.
14. Morge S, Adamczak E, Lindèn LA. Variation in human salivary pellicle formation on biomaterials during the day. Arch Oral Biol. 1989;34:669–74.
15. Cavalcanti YW, Wilson M, Lewis M, Williams D, Senna PM, Del-Bel-Cury AA, da Silva WJ. Salivary pellicles equalize surfaces' charges and modulate the virulence of Candida albicans biofilm. Arch Oral Biol. 2016;66:129–40.
16. Jendresen MD, Glantz PO. Clinical adhesiveness of selected dental materials. An in-vivo study. Acta Odontol Scand. 1981;39:39–45.
17. Cassiano LPS, Ventura TMS, Silva CMS, Leite AL, Magalhaes AC, Pessan JP, Buzalaf MAR. Protein profile of the acquired enamel pellicle after rinsing with whole milk, fat-free milk, and water: an in vivo study. Caries Res. 2018;52:288–96.
18. Delius J, Trautmann S, Médard G, Kuster B, Hannig M, Hofmann T. Label-free quantitative proteome analysis of the surface-bound salivary pellicle. Colloids Surf B Biointerfaces. 2017;152:68–76.
19. Lee YH, Zimmerman JN, Custodio W, Xiao Y, Basiri T, Hatibovic-Kofman S, Siqueira WL. Proteomic evaluation of acquired enamel pellicle during in vivo formation. PLoS One. 2013;8:e67919.
20. Taira EA, Ventura TMS, Cassiano LPS, Silva CMS, Martini T, Leite AL, Rios D, Magalaaes AC, Buzalaf MAR. Changes in the proteomic profile of acquired enamel pellicles as a function of their time of formation and hydrochloric acid exposure. Caries Res. 2018;52:367–77.
21. Ventura TMSD, Cassiano LPS, de Souza e Silva CM, Taira EA, Leite AL, Rios D, Buzalaf MAR. The proteomic profile of the acquired enamel pellicle according to its location in the dental arches. Arch Oral Biol. 2017;79:20–9.
22. Schweigel H, Wicht M, Schendicke F. Salivary and pellicle proteome: a datamining analysis. Sci Rep. 2016;6:38882.
23. Trautmann S, Barghash A, Fecher-Trost C, Schalkowsky P, Hannig C, Kirsch J, Rupf S, Keller A, Helms V, Hannig M. Proteomic analysis of the initial oral pellicle in caries-active and caries-free individuals. Proteom Clin Appl. 2019;13:e1800143.
24. Hannig M. Transmission electron microscopic study of in vivo pellicle formation on dental restorative materials. Eur J Oral Sci. 1997;105:422–33.
25. Hannig M. Elektronenmikroskopische Untersuchungen der initialen Bioadhäsionsprozesse an Festköperoberflächen in der Mundhöhle. Berlin: Quintessenz Verlags-GmbH; 1998.
26. Song F, Koo H, Ren D. Effects of material properties on bacterial adhesion and biofilm formation. J Dent Res. 2015;94:1027–34.
27. Rios D, Honório HM, Francisconi LF, Magalhães AC, de Andrade Moreira Machdo MA, Buzalaf MAR. In situ effect of an erosion challenge on different restorative materials and on enamel adjacent to these materials. J Dent. 2008;36:152–7.
28. Bilhan H, Bilgin T, Cakir AF, Yuksel B, von Fraunhofer JA. The effect of mucine, IgA, urea, and lysozyme on the corrosion behavior of various non-precious dental alloys and pure titanium in artificial saliva. J Biol Appl. 2007;22:197–221.
29. Clark GC, Williams DF. The effects of proteins on metallic corrosion. J Biomed Mater Res. 1982;16:125–34.
30. Mezger PR, van't Hof MA, Vrijhoel MA, Gravenmade EJ, Greener EH. Effect of mucine on the corrosion behavior of dental casting alloys. J Oral Rehabil. 1989;16:589–96.
31. Bozzini B, Carlino P, D'Urzo L, Pepe V, Mele C, Venturo F. An electrochemical impedance investigation of the behavior of anodically oxidized titanium in human plasma and cognate fluids, relevant to dental applications. J Mater Sci Mater Med. 2008;19:3443–53.
32. Höhn S, Braem A, Neirinck B, Virtanen S. Albumin coatings by alternating current electrophoretic deposition for improving corrosion resistance and bioactivity of titanium implants. Mater Sci Eng C Mater Biol Appl. 2017;73:789–807.
33. Finkelstein GF, Greener EH. In vitro polarization of dental amalgam in human saliva. J Oral Rehabil. 1977;4:347–54.
34. Håkansson B, Yontchev E, Vannerber NG, Hedegard B. An examination of the surface corrosion state of dental fillings and constructions. I. A laboratory investigation of the corrosion behavior of dental alloys in natural saliva and saline solutions. J Oral Rehabil. 1986;13:235–46.
35. Shinobu K. The anticorrosive action of saliva against metals. J Jpn Res Soc Dent Mater Appl. 1955;2:42.
36. Holland RI. Effect of pellicle on galvanic corrosion of amalgam. Scand J Dent Res. 1984;92:93–6.
37. Hahn Berg IC, Rutland MW, Arnebrant T. Lubricating properties of the initial salivary pellicle—an AFM study. Biofouling. 2003;19:365–9.
38. Joiner A, Schwarz A, Philpotts C, Cox TF, Huber K, Hannig M. The protective nature of pellicle towards toothpaste abrasion on enamel and dentin. J Dent. 2008;36:360–8.
39. Zhang YF, Zheng J, Zheng L, Shi XY, Zhou QZR. Effect of adsorption time on the lubricating properties of the salivary pellicle on human tooth enamel. Wear. 2013;301:300–7.
40. Zhang YF, Zheng J, Zheng L, Zhou ZR. Effect of adsorption time on the adhesion strength between sal-

ivary pellicle and human tooth enamel. J Mech Behav Biomed Mater. 2015;42:257–66.
41. Hahn Berg IC, Lindh L, Arnebrant T. Intraoral lubrication of PRP-1, statherin and mucin as studied by AFM. Biofouling. 2004;20:65–70.
42. Zhang YF, Li DY, Yu JX, He HT. On the thickness and nanomechanical properties of salivary pellicle formed on tooth enamel. J Dent. 2016;55:99–104.
43. Bordin D, Cavalcanti IMG, Pimentel MJ, Fortulan CA, Sotto-Maior BS, Del Bel Cury AA, da Silva WJ. Biofilm and saliva affect the biomechanical behavior of dental implants. J Biomech. 2015;48:997–1002.
44. Cavalcanti YW, da Silva Girundi FM, Assis MAL, Zenobio EG, Soares RV. Titanium surface roughing treatments contribute to higher interaction with salivary proteins MG2 and lactoferrin. J Contemp Dent Pract. 2015;16:141–6.
45. Mei L, Busscher HJ, van der Mei H, Ren Y. Influence of surface roughness on streptococcal adhesion forces to composite resins. Dent Mater. 2011;27:770–8.
46. Al-Ahmad A, Wiedemann-Al-Ahmad M, Fackler A, Follo M, Hellwig E, Bächle M, Hannig C, Wolkewitz M, Kohal R. In vivo study of the initial bacterial adhesion on different implant materials. Arch Oral Biol. 2013;58:1139–42.
47. Park JW, Song CW, Jung JH, Ahn SJ, Ferracane JL. The effect of surface roughness of composite resin on biofilm formation of *Streptococcus mutans* in the presence of saliva. Oper Dent. 2012;37:532–9.
48. Ionescu AC, Cazzaniga G, Ottobelli M, Ferracane JL, Paolone G, Brambilla E. In vitro biofilm formation on resin-based composites cured under different surface conditions. J Dent. 2018;77:78–86.
49. Swartjes JJTM, Sharma PK, van Kooten TG, van der Mei HC, Mahmoudi M, Busscher HJ, Rochford ETJ. Current developments in antimicrobial surface coating for biomedical applications. Curr Med Chem. 2015;22:2116–29.
50. Imazato S, Ebi N, Takahashi Y, Kaneko T, Ebisu S, Russell RRB. Antibacterial activity of bactericide-immobilized filler for resin-based restoratives. Biomaterials. 2003;24:3605–9.
51. Li F, Weir MD, Fouad AF, Xu HHK. Effect of salivary pellicle on antibacterial activity of novel antibacterial dental adhesives using a dental plaque microcosm biofilm model. Dent Mater. 2014;30:182–91.
52. Müller R, Eidt A, Hiller KA, Katzur V, Subat M, Scheikl H, Imazato S, Ruhl S, Schmalz G. Influences of protein films on antibacterial or bacteria-repellent surface coatings in a model system using silicon wafers. Biomaterials. 2009;30:4921–9.
53. Bell A, Creanor SL, Foye RH, Saunders WP. The effect of saliva on fluoride release by a glass-ionomer filling material. J Oral Rehabil. 1999;26:407–12.

3 Oral Biofilms and Secondary Caries Formation

Eugenio Brambilla and Andrei Cristian Ionescu

Abstract

The presence of a huge amount of data regarding the onset of a carious lesion in close proximity with a restoration must not make ourselves less aware about the fact that, still, a lot of information is missing about secondary caries formation. Many pieces of information are highly conflicting, such as the role that different microbial species have in the onset of the lesion, or the link between the existence—and width—of a gap between hard tissues and restoration, and the development of a secondary lesion, or, again, the clinical decisions regarding the replacement of a restoration due to secondary caries, and, if so, to what point stop excavating.

The main difficulty in this field arises from the fact that secondary caries is the result of very complex interactions taking place among already injured human tissues, overlying biofilms that often maintain the dysbiotic conditions that had lead to the primitive lesion, and dental (bio) materials that may help, or even worsen that situation. Increasing our knowledge about what exactly happens after a material is placed at this three-sided interface may greatly help us in designing new dental materials able to interact in a positive way both with the host and its biofilm, ensuring longevity to restorations and helping in reducing what is nowadays their main cause of failure—secondary caries.

E. Brambilla · A. C. Ionescu (✉)
Oral Microbiology and Biomaterials Laboratory, Department of Biomedical, Surgical, and Dental Sciences, University of Milan, Milan, Italy
e-mail: eugenio.brambilla@unimi.it; andrei.ionescu@unimi.it

3.1 Secondary Caries: What Is It?

The answer is -we still do not know for sure. In 2009 Cenci et al. stated: "Up to now, there has been no conclusive evidence for the association or lack of association between gap presence and caries adjacent to restorations." After 10 years, this sentence still seems to be valid. However, we must start from some definition [1].

Secondary or recurrent caries is a lesion at the marginal area of an existing restoration [2].

Contrarily to residual caries, which is represented by infected tissue left behind after cavity preparation, secondary caries develops close to a restoration. Nevertheless, in clinical practice, it is most often difficult to differentiate between these situations. This is the reason why secondary caries have been receiving increasing attention over the last years. It has become a matter of concern in restorative dentistry since it has been recognized as the most common reason for premature failure of restorations, irrespective of the restorative material used [3–5].

A. C. Ionescu, S. Hahnel (eds.), *Oral Biofilms and Modern Dental Materials*,
https://doi.org/10.1007/978-3-030-67388-8_3

This fact is demonstrated by a relevant amount of experimental data about the clinical performance of dental restorative materials that have been published over the last two decades [3, 6–9]. In these studies, the development of a secondary caries lesion is considered one of the most important parameters to measure the performances of the restorative materials. Furthermore, recent literature indicates that secondary caries and its prevention is one of the critical problems of the next 20 years [10].

For many reasons, epidemiological data about the prevalence and the incidence of secondary caries are far from being complete and exhaustively investigated. Firstly, there are no consensual standards to perform the detection of secondary caries. Clinicians use many empirical methods based on their field experience. This approach leads to relevant differences in decision-making regarding the need for replacement of the restorations, which might be grounded on false-positive diagnoses that may finally cause unnecessary replacement of restorations [5, 6]. This considerable heterogeneity of diagnosis-subsequent treatment decisions is based on criteria with limited accuracy [11–14]. From an operative point of view, the lack of a clear and shared strategy causes an unmotivated sacrifice of sound tooth tissues and coincides with unnecessary costs and a decrease of the functional shelf-life of natural teeth [15]. Citing Elderton, decisions made by dentists for replacing a filling are idiosyncratic [16]. This circumstance may explain not only the high variability that is reported in the literature regarding the incidence of secondary caries but to a certain extent also the higher incidence of secondary caries that has been observed in practice-based studies that included a higher number of noncalibrated operators than academic studies [17, 18]. On the other hand, the prevalence and incidence of secondary caries identified in controlled clinical trials may not be representative of daily dental practice. As such, the current approach in addressing secondary caries treatment has a profound impact on healthcare expenses, since the replacement of existing restorations due to secondary caries takes most of the work time of the dentist [19–21].

An estimated 60% of the replaced resin-based composite restorations, however, belong to a "redentistry cycle" which describes a continued series of repeated restoration placements, implying that the patient is supplied with more restorations depending on the existing restorations [16]. A clinical example is given (Fig. 3.1).

Moreover, subsequent replacement or repair of a restoration leads to further loss of natural tooth tissues, starting a so-called death spiral, which may eventually lead to tooth loss and reduction of the residual oral health [17, 18].

Secondary caries can be considered the same disease as primary caries, and patient-related factors such as oral hygiene and dietary habits play a crucial role in its etiology. Nonetheless, there is also evidence that the problem is related to the interface between natural tooth tissues and the restorative material as well as the characteristics of the restorative material [4]. Consequently, increasing efforts are made to modulate the characteristics of such interface in order to improve the ability of restorative materials to reduce the onset of secondary caries [22]. However, the detection of secondary lesions has received less attention until now, even though it is essential both to estimate the incidence of the disease and to manage it effectively. In fact, this problem still seems to be unsolved.

Many clinical studies showed a higher incidence of secondary caries associated with resin-based composites than with amalgam, which implies that resin-based composites could be more susceptible to secondary caries than amalgams especially in high-caries-risk subjects [20, 23, 24]. The higher susceptibility of resin-based composites to secondary caries has so far been associated with various material-inherent properties such as polymerization shrinkage and subsequent microleakage, higher plaque accumulation,

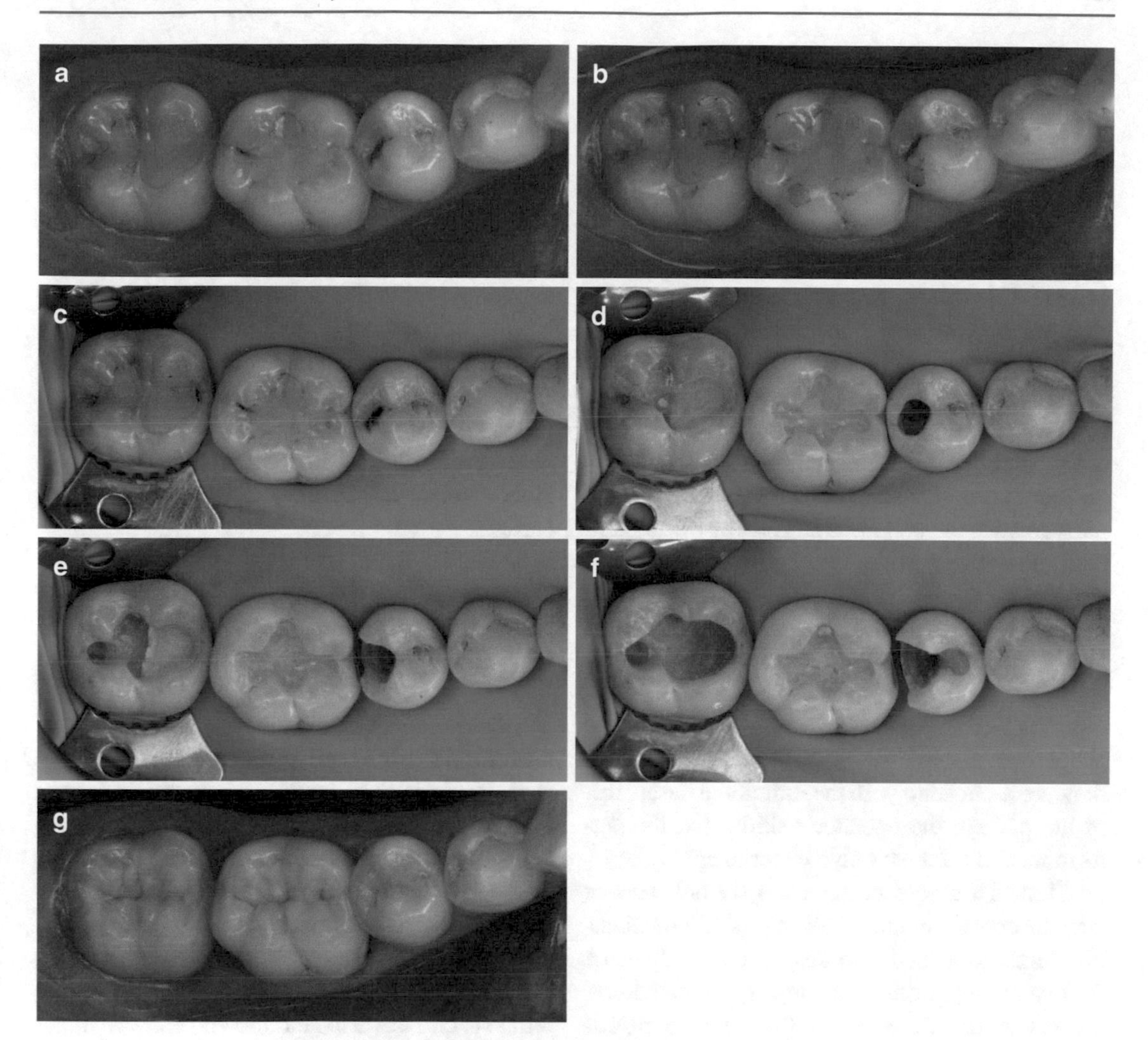

Fig. 3.1 A clinical case is shown where restorations were performed or replaced due to primary and secondary caries. The diagnosis step is crucial and can never be underestimated (**a**). It can be seen that the lesion considered as secondary caries on the lower first molar was, in fact, a superficial stain (**d**). The latter could have easily been removed by finishing procedures applied to the marginal area of the restoration. The choice of replacing the restoration of the lower first molar was, in this case, only motivated by occlusal reasons (**b**). Nevertheless, it must be highlighted that a similar result could have probably been achieved in a more conservative way by a reshaping of the restoration performed with a finishing bur. On the contrary, relatively minor signs on the marginal area of the lower second molar restoration (**c**) hide the presence of deep secondary caries extended beyond the floor of the original restoration (**e**, **f**). The lower second premolar also showed that deep caries originated both from the occlusal and interproximal surfaces (**c**, **e**). Re-performing restorations always leads to bigger cavities and, consequently, to reduced residual health of the tooth. The excavation was limited to reaching firm tissue with a leathery consistency after checking that restorations' margins were made of sound tissues ensuring a perfect seal. The completed restorations have to reproduce the tooth function, aesthetics, and, possibly, biological behavior (**g**). (All pictures were acquired during experimental or clinical activity performed by the authors of this chapter)

release of bacterial growth-stimulating compounds, lack of antibacterial and acid-buffering effects, and changes in microbial composition [4, 25, 26]. Thus, several studies were performed, aiming to improve the behavior of resin-based composites in order to make the restorations more resistant to the development of secondary caries. The main strategy is represented by changes in chemical composition produced by the addition of antimicrobial compounds, either to the formulation of the resin-based composites or to the corresponding adhesive, providing generally satisfying results in vitro in the short-middle term [27–30]. Clinical data on the performance of such materials are, however, generally lacking. It is challenging to design studies that produce relevant and compelling data about the progress of initial secondary caries, which is due to ethical reasons and the intrinsic difficulties associated with clinical trials, such as their duration. Therefore, the identification of potential prognostic factors is difficult, mainly because they are associated with the individual caries risk of the patient, the operator's skills, and the performance of the restorative material applied.

Clinical trials are, nevertheless, the only current way to convey reliable indications to clinicians. Problems associated with this kind of study occur for two main reasons, including (1) the high number as well as different types of restorative materials available on the market and (2) the specific application techniques. Moreover, restorative materials have one of the shortest turnover times. In many cases, restorative materials investigated in a clinical study have already been replaced by the next-generation materials when the study is finished, meaning that the results are outdated before the completion of the study. Moreover, results and conclusions drawn by analyzing a particular type of restorative material can hardly be transferred to the behavior of other restorative materials even from the same class [31].

From this point of view, the duration of clinical studies produces other critical limitations. Planning, performing, and reporting the outcomes of a clinical trial on secondary caries are challenging as there may be high dropout rates of the participants, and manufacturers often ask for experimental designs involving observation periods that are less than 3 years. However, the favorable clinical performance of a restorative material over short periods is no predictor for favorable long-term behavior. Thus, we need a strong development in the field of accelerated artificial aging and microbiological aging, that is just beginning to be considered. Some of the models currently applied to study materials-tissue-biofilm interactions at the interface are explained in the next Chapter. In any case, alternative strategies are being developed, and expected to improve the performance of new generations of restorative materials in order to protect the marginal area of surrounding tissues by secondary caries.

3.2 Microbiology of Secondary Caries

The development of a caries lesion is caused by an imbalance between pathological factors that lead to a loss of minerals and protective factors that cause an uptake of ions by the tooth tissues [32]. Cariogenic bacterial species represent the leading etiologic agent for this process. Their fermentative metabolism produces organic acids, such as carbonic, lactic, and propionic acid. It has been known for years that those acidogenic bacterial species are also aciduric and can live under acidic conditions [33]. The microbiology of secondary caries is even more complicated due to the presence of the restorative material that unevenly interacts with the biofilm colonizing both the surface and the interface.

Although the histopathology of secondary caries is described as similar to that of primary caries, its etiology is not as clear. Beginning in the 1990s, research efforts in this field have until now failed to produce consistent data. Kidd et al. showed no significant differences between the microflora in samples from cavity walls involving primary caries and secondary caries in the proximity of amalgam restorations [34]. Thomas et al., using an in situ model to investigate the composition of biofilms colonizing the surfaces in primary and secondary caries lesions, identified a higher proportion of cariogenic bacteria on restorations fabricated from resin-based composite. These data show that the composition of biofilms in primary caries lesions differs from that of secondary lesions developing in the proximity of resin-based composite restorations [35].

Another complex topic is the colonization of tissues under the restorations. Mejàre and Malmgren found that the bacterial flora below resin-based composite restorations is similar to the flora observed in dental plaque, mainly including *Streptococcus* and *Actinomyces* spp. [36]. Splieth et al. [37] compared the microbial flora under resin-based composite and amalgam restorations focusing on the anaerobic species, identifying a similar bacterial composition under both materials. In particular, the latter study indicated that inadequate resin-based composite restorations might stimulate the growth of cariogenic as well as obligate anaerobic bacteria potentially pathogenic to the pulp. There are many possible explanations for these observations. Firstly, the micro space between the restoration and the cavity floor favors the ecological niche of obligately anaerobic bacteria. In fact, it is not surprising to discover many obligate anaerobic bacteria commonly colonizing the oral environment, even in subjects without clinically detectable endodontic or periodontal lesions. In addition to that, it does not necessarily mean that subjects without clinical symptoms such as toothache or pulpitis do not have chronic or arrested caries lesions, which implies that it is still possible to detect such anaerobic bacteria. However, it must be highlighted that coexistence does not mean involvement. Therefore, it is still necessary to demonstrate the participation of those obligate anaerobic bacteria in the progress of secondary caries. These considerations suggest that the restorative material plays a crucial role in the composition of the biofilms colonizing surfaces and interfaces of dental restorations and surrounding tooth tissues.

According to the viewpoint of Philip D. Marsh, any species with an acidogenic ability that is able to tolerate the cariogenic environment can contribute to the progress of dental caries [38]. For a long time, mutans streptococci (*S. mutans*), lactobacilli, and *Actinomyces naeslundii* have been used in several in vitro models to study secondary caries. *S. mutans* and lactobacilli can produce a variety of organic acids and can withstand a low pH environment for a long time, thus leading to the demineralization of dental tissues and the onset of caries. It has been shown that these bacterial species are widely present and might play a crucial role in the development of secondary caries around amalgam restorations [39]. However, in a recent in situ study, *S. mutans* was not detected in any sample, contrarily to lactobacilli spp. Furthermore, *Actinomyces odontolyticus* and *Candida* spp. were also found in most samples [35]. In fact, David Beighton recently put forward a distinct point of view, insisting that *S. mutans* might be a useful marker of secondary caries without necessarily being its etiological agent [40]. These assumptions are supported by the previously described experiments performed by Renske Thomas's research group.

As a consequence, scientists speculated that there might exist unknown caries-associated bacteria that cannot be isolated on selective agar plates [35]. In the past decade, the detection of *A. odontolyticus* and *Candida* spp. associated with caries development has caused reactions of surprise in the scientific community. It has been shown that *Candida albicans*, despite its lower growth rate [41], can dissolve hydroxyapatite at a much higher rate than *S. mutans*.

Klinke et al. assumed that *C. albicans* might make a significant contribution to caries pathogenesis in caries-active children, and it could be responsible for an increase in caries pathogenicity [42]. Some experiments performed by the authors of this chapter show signs of increased virulence (enhanced adherence to the pellicle-treated enamel surfaces, increased replication rate, and extracellular matrix production) by *S. mutans* when co-cultured with the yeast, while none of these microorganisms seemed to suffer from the presence of their "partner," possibly meaning that they express synergistic relationships (Fig. 3.2).

Apart from that, it should be noted that many subjects may have high caries activities without having a prevalence of *S. mutans* in the composition of oral biofilms. Therefore, further research is needed to define better the microbiological features of secondary caries, such as the role of the

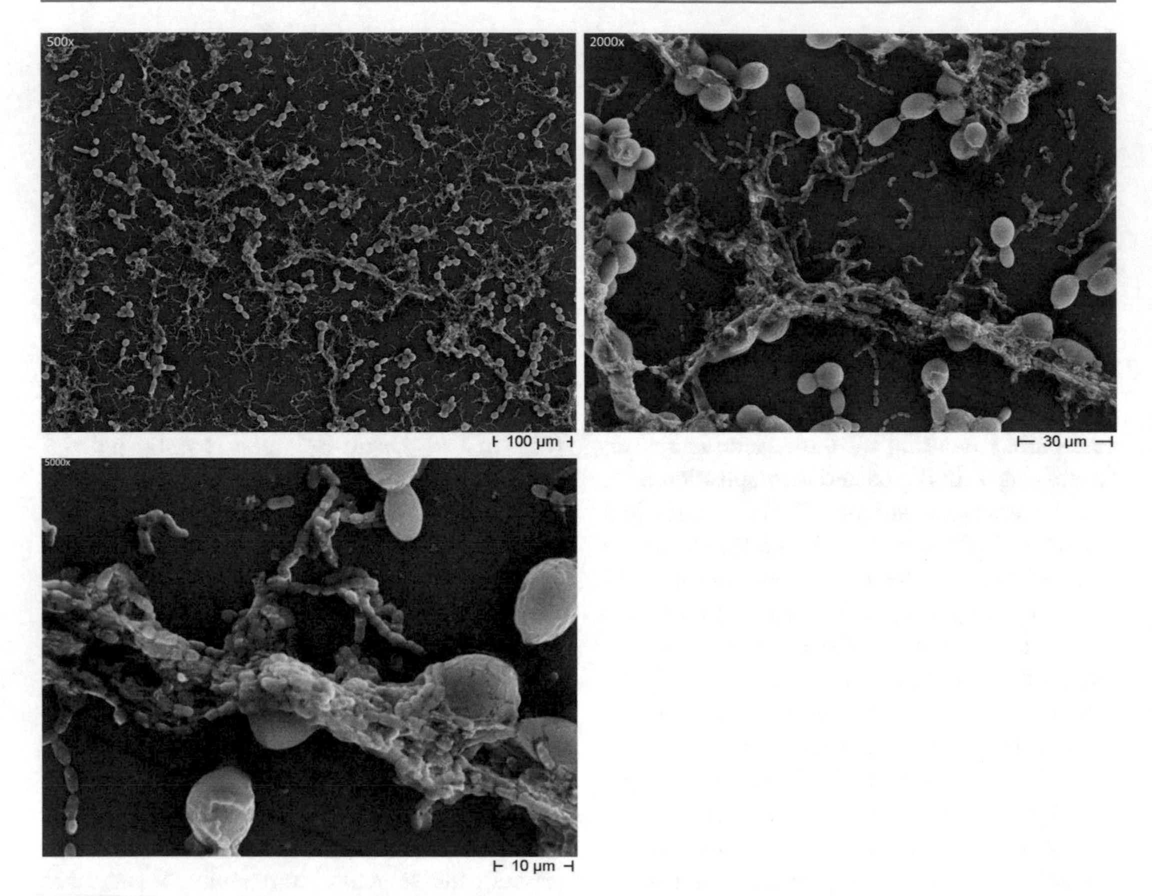

Fig. 3.2 A series of SEM micrographs showing a co-culture of *S. mutans* and *C. albicans* at increasing magnifications. The microorganisms were cultivated in a bioreactor, starting from a 1:1 inoculum using a sucrose-enriched salivary mucin medium over enamel surfaces for 24 h. The adherence surfaces were pretreated with sterile human saliva for 24 h before inoculation to allow the formation of a salivary pellicle. Signs of active growth can be seen for both microorganisms, such as the formation of streptococcal chains showing several microorganisms replicating at the same time, and budding of a high number of yeast blastospores. At this timepoint, *S. mutans* outcompetes *C. albicans* in terms of cell numbers, and the pH near the surface is less than 4.5. Despite that, neither microorganism shows signs of suffering. More interestingly, *S. mutans* preferentially adheres to *C. albicans* cells and more promptly produces extracellular matrix than when directly adhering to enamel surfaces, suggesting that the yeast improves the virulence of the bacterium. At the same time, the latter's activity is not detrimental to the yeast, and the extracellular matrix produced by the coccus may serve as protection for both microorganisms

different pathogenic species and interspecies relationships.

3.3 Carious Tissues Features and Removal Strategies of Secondary Caries

In a secondary lesion, we can identify two distinct regions: the surface lesion, which develops perpendicular to the tooth surface and can be considered as a primary lesion developing next to a restoration, and the wall lesion, which develops in-depth, perpendicular to the tooth/restoration interface (Fig. 3.3 [43]).

Histological analysis of artificial, caries-like lesions and natural lesions around restorations may yield lines of demineralized tissue running along the cavity wall. These are called wall lesions, are considered to be the result of microleakage, and can be identified in natural teeth with occlusal amalgam restorations. In this case, the wall lesions are probably the consequence of an initial leakage that occurred just after the

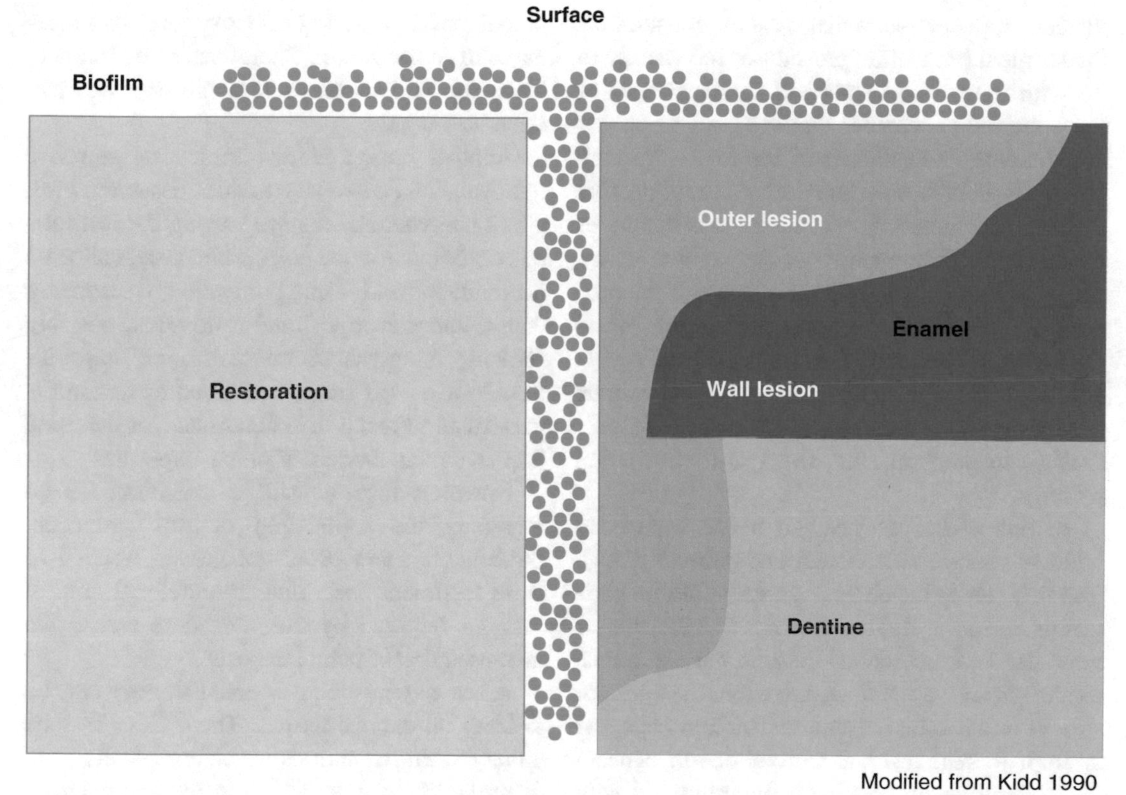

Fig. 3.3 In a secondary lesion, two distinct regions exist, the outer lesion and the wall lesion. The first one can be considered as a primary lesion developing next to a restoration, while the wall lesion develops perpendicular to the tooth/restoration interface and is the result of microleakage

placement of the restoration and before the filling of the marginal gap with corrosion products.

Particular attention must be paid not to mix up secondary caries with histological signs of microleakage. The difficulty to clinically distinguish between these situations is a crucial factor in making a clinical decision for replacement of restorations, which often causes overtreatment. Furthermore, it is also important not to mix up secondary caries with residual caries, which is to be regarded as residual demineralized tissue that is more or less intentionally left during cavity preparation. This distinction is a very difficult task for both clinicians and researchers because the clinician needs to know how much carious dentin should be removed and how deep the excavation has to be extended [44, 45]. The criteria defining the amount of excavation and caries removal are influenced both by the interaction of the surgical procedure itself with pulp tissues and by the application of the adhesive procedures. Clinical studies of Mertz-Fairhurst have questioned the consolidated approach to how much demineralized tissue may be left during cavity preparation [46]. In their clinical studies, the enamel lid from large occlusal lesions was removed, leaving extensively demineralized dentin. The cavities were then restored with adhesive material techniques. After 10 years, data very surprisingly showed that these restorations were still satisfactory, with no need for replacement. These results put into question the conventional teaching in restorative dentistry. Indeed, we had been assuming that the infected, demineralized dentin that is part of the carious lesion must be removed entirely in order to arrest the progression of a carious lesion.

Nevertheless, if no negative effect happened after leaving the infected tissue on the bottom of the cavity, the endpoint that we must consider for the surgical excavation procedures has mainly to do with the requirements and characteristics of the adhesive procedures that are used to restore the function and esthetics of the tooth. We may better explain these results after accepting that the bacterial metabolism of biofilms is the driver for caries development and progression at all stages of this pathology. In this sense, if the process is arrested by removing the biofilm inside the lesion, or, better said, by tuning down its metabolic activity, the entity of the remaining carious tissue removal is merely a function of the tissue's ability to support the overlying restoration reliably.

In this sense, we need to define criteria to address restorative interventions rationally [44]. The first step is the choice of a threshold for carious tissue removal. This choice is still a matter of huge debate in the dental field. In general terms, clinicians and researchers agree that the operator should retain sound, remineralizable tissues and attempt to seal residual carious dentin beneath the restoration by creating a durable bond with sound tooth tissues surrounding the lesion (Fig. 3.1). Another crucial target is to save the vitality of the pulpal tissues by avoiding as much as possible accidental exposures.

Schwendicke et al. summarized five main strategies to maximize the success of the removal of carious tooth tissues. Following these indications, our operative intervention can span from the drastic removal of all the softened dentin to completely avoiding this step. The real focus of the discussion has been, regardless of the excavation strategy, that the remaining tissues still harbor vital bacteria. The new approach is based on the idea that the removal of all microorganisms in the tissue is not quite necessary [44].

From this point of view, the strategy involving nonselective removal to hard dentine (formerly complete excavation or complete caries removal) is now considered overtreatment and no longer recommended. Less invasive and consequently more selective removal to firm dentine leaving a bottom layer of leathery tissue and sound margins is the treatment of choice in carious lesions that are not in close proximity with the pulp. Deeper lesions produce a significant risk of pulpal exposure during selective removal procedures to sound dentin. Therefore, an option is to leave soft carious tissue to avoid exposure and stress to the pulp.

Another option is represented by stepwise removal, that is, two-step carious tissue removal: a first intervention involves leaving the soft carious tissue close to the pulp, while time is allowed for remineralization and production of secondary dentin under a provisional restoration, possibly showing bioactivity/remineralization capabilities. In a second stage, additional excavation is performed to reach firm dentin and, at the same time, reducing the risk of pulpal exposure.

However, there is still a consistent debate regarding the application of this technique. Indeed, the main disadvantages are related to extra treatment costs due to additional clinical sessions required by this procedure and to the increased risk of pulpal exposure.

A last and most controversial strategy may be to leave all carious tissues. The options include either to seal the carious lesion completely or to open the lesion to expose it to the oral environment, and to manage it without further restorative procedures. Sealing of non-cavitated or minimally cavitated lesions in areas not subjected to occlusal load was demonstrated to be a highly effective option. The second possibility is based on the induction of massive changes in the composition of the biofilm that colonizes the superficial layers of the carious tissues and its pathogenicity. Once a lesion is exposed to the oral environment, the increased interaction with saliva and its components dramatically reduces the lesion's activity. This possibility, however, requires very high compliance from both the patient and the operator, involving frequent recalls and follow-up and, at the bottom line, was not found to be more effective than its alternatives.

3.4 Secondary Caries and Microleakage

The presence of a not perfectly sealed interface between the restorative material and the surrounding hard tissues has been regarded as the main reason for the development of secondary caries for many years. The reasons were assumed to be microbial penetration into a gap between the restorative material and the surrounding sound dental tissues and production of acidic metabolites in this new microenvironment [43, 47, 48].

One of the possible situations leading to the development of both outer and wall lesions is the existence of the gap itself. The size of this gap can vary from a few microns to some hundred microns [49–51]. The presence of a clinically visible gap is regarded as a sign of the presence of defective restoration margins and often leads to the decision to replace the restoration [52]. Therefore, the role of this structure seems to be crucial for clinical decision-making and has to be carefully considered to avoid overtreatment.

Two explanations attempt to put into relation gaps with the development of a cavity wall lesion. The first one postulates the penetration of bacterial cells and, especially, their metabolites through the gap (microleakage), thus initiating the demineralization process. According to this "microleakage theory," bacterial colonization increases along with the size of the gap [26, 53, 54]. Nevertheless, this microleakage-based theory has been questioned recently by research data indicating that secondary caries is but a primary lesion that develops in the marginal area of a restoration. The evolution of a lesion at the restoration margin is determined by the activity of the biofilm colonizing the outer surface—regardless, within reasonable limits, of the presence of microleakage. The cavity wall lesion is, therefore, a consequence of the extension of the process already taking place on the external surface. Furthermore, the microleakage-based theory does not take into account that biofilm development is dependent on environmental parameters such as the oxygen diffusion gradient or chemical compounds leaching from the material's surface. In a tiny gap, these variables may not necessarily be more favorable than in a more substantial gap.

The second explanation is based on the recent advance of knowledge regarding the microbiology of cariogenic biofilms and their role in secondary caries development. Both in vitro and clinical experimental data showed that microleakage alone does not necessarily promote an active demineralization process on the cavity wall of a restoration [55]. Bacterial colonization of the tooth-restoration interface is a mandatory prerequisite for the development of a secondary carious lesion, just as in primary caries.

Nevertheless, there are relatively few experimental data in the literature dealing with the relationships between gap size and wall lesion development. Indeed, we still do not know if a minimum gap size can be determined to cause the lesion development. Moreover, even if this information were available, its use by the clinician in the decision-making process on the replacement of a restoration or parts of it would be tough. Data about the relationship between gap presence and secondary caries development is still controversial [1]. Some studies identified a lack of correlation between these factors [56–58], while other researchers have found a positive relationship based on gap size [54, 59].

In particular, regarding the results gathered by the most recent in vitro studies, Totiam et al. and Nassar and González-Cabezas used different sucrose-cycling *S. mutans* models, showing that in experimental gaps the size of the gap is positively correlated with the size of dentinal wall lesions [53, 54]. Only Diercke et al. [60] demonstrated a statistically significant increase in lesion depth in enamel (50–250 μm gap) and dentin (50–100 μm gap). It must be pointed out that in this study, as in the one by Nassar and González-Cabezas, no adhesive system was used, so probably the microenvironment was very different from the clinical situation. Furthermore, the experimental design of these studies shows other bias sources related to a relatively short incubation period, the use of a static setup (no bioreactor to simulate biofilm formation under shear stresses), and the use of monospecies *S. mutans* biofilm. For these reasons, the results above have to be interpreted with caution.

In recent years, the improvement of microbiological techniques and the diffusion of the use of bioreactors seem to be improving the reliability and translational value of results obtained by in vitro studies (see also Chap. 4). The correlation between a gap and secondary caries is confirmed by the results of Hayati et al. [61] that—using a bioreactor—studied the gap colonization by a multispecies cariogenic biofilm. The influence of an adhesive system was also evaluated. The results showed that the presence of an adhesive system significantly reduced the progression of the secondary lesion. These findings show that the results of the previously described studies might be overestimated due to the absence of an adhesive system. Finally, Maske et al. [62] investigated the development of dentin wall lesions next to resin composite using an in vitro microcosm model. Their aim was to evaluate the influence of the gap size on the wall lesion development. They found that wall lesions in dentin developed even in tiny gaps, and the threshold for secondary wall lesion development was around 30 μm.

Regarding in situ studies, Thomas et al. [51] indicated that an average gap size of 225 μm was necessary for the development of a wall lesion, even if their results had a broad data range (80–560 μm). In 2009, Lima et al. demonstrated a significant impact of biofilm control in the prevention of enamel lesion demineralization. Their results suggest that microleakage and surface roughness do not influence the formation of secondary caries lesions. In 2015, Montagner et al. demonstrated that the presence of a bonding agent on the composite side of a restoration–dentin gap increases wall lesion development [63]. One year later, the same group demonstrated that composite-dentin interfaces that failed after aging showed different demineralization patterns depending on the presence of an adhesive system [64]. These data showed that the restorative procedure and the application of the adhesive system deeply influence the response of the tissue structure to secondary caries challenge (Fig. 3.4).

In conclusion, it is interesting to note that the integrity of the composite-tissue interface may be of critical importance for the development of secondary caries in adhesive restorations.

3.5 Is Secondary Caries a Material-Related Problem?

There is currently a broad spectrum of restorative materials available from which clinicians can choose. Each type has its own physical and chemical characteristics that influence its field of application. As a consequence, we have an equally broad spectrum of interfaces between dental tissues and restorative materials. It is, therefore, challenging to obtain univocal results in the study of secondary caries development. During the past decade, restorative materials underwent a paradigm shift, changing from amalgam to adhesive materials. This development caused significant improvements in the esthetic performance and posed the basis to propagate minimally invasive surgical techniques. Resin-based composites have become the most commonly used restorative material [65, 66], and they have gradually replaced the amalgam since the latter has been associated with mercury toxicity (not yet demonstrated) and environmental problems [67, 68].

Furthermore, their bonding ability to tooth tissues allows a wide variety of potential applications. For instance, direct treatment of clinical situations that had previously to be treated with indirect techniques is now possible. Composites also represent the ideal interface for bonding ceramic materials to tooth tissues, allowing a level of performance unknown until recently. One of the fields in which these materials have allowed for most significant developments is the minimally invasive approach. Adhesive techniques made it possible to remove only the necessary amount of dental tissues, thus minimizing the sound tissue sacrifice [69, 70].

Speaking about interfaces from a material's point of view, the terms used for any type of restorative material are "margin," "adaptation," or "gap." Other terms, for example, "marginal seal or sealing," are primarily associated with the adhesive interface of a restorative material. "Ditch" or "flowing" is mainly used for amalgams. The differences among the terms used represent the existence of a broad spectrum of situations that are related to different chemical and physical properties of the restorative materials. Moreover, the interfaces are dynamic microenvironments that show a complex balance based on the exchange of chemical compounds leaching from the restorative material and ions and other compounds from the external environment (saliva) and the dental tissues.

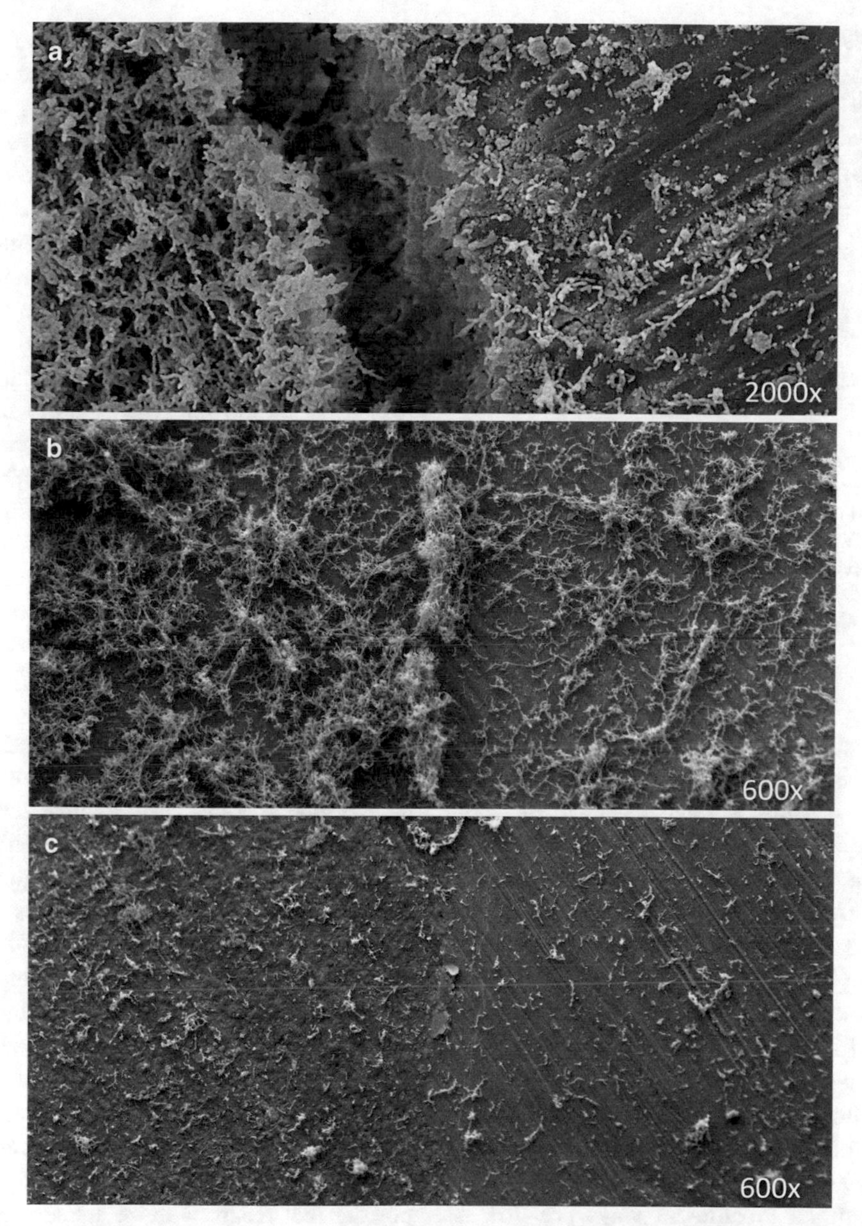

Fig. 3.4 A series of SEM micrographs are shown of an aerial view of an enamel-composite interface. In all micrographs, the enamel is on the right, and the composite restoration surface is on the left, while the interface is displayed vertically. (**a**) A composite restoration was performed on an enamel slab, and in vitro, *S. mutans* biofilm formation was obtained using a bioreactor for 48 h. The microorganism's colonization can be predominantly seen on the composite surface and in the microgap between the materials (2000× magnification). (**b, c**) The enamel-composite interfaces are shown where different adhesive systems were used to bond composite restorations to enamel slabs (600×). In (**b**), a conventional self-etch adhesive system was applied, while in (**c**), a self-etch adhesive featuring an antibacterial monomer (MDPB) was used. In the first case, after in vitro *S. mutans* biofilm formation, intense colonization of the composite surface and the adhesive interface can be seen. On the contrary, the antibacterial adhesive system was able to reduce bacterial colonization not only at the interface but also on the surfaces in close proximity. No evidence of gap can be seen, suggesting that both adhesives performed optimal sealing of the interface and that microbial degree of colonization of the interface was not dependent on the presence of a gap

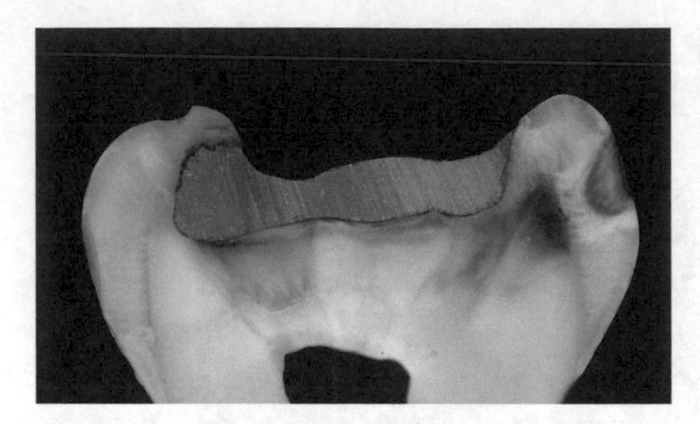

Fig. 3.5 Section of an extracted lower molar with an amalgam restoration. Clinical appearance does not reflect the condition of the underlying tissues, not affected by a secondary lesion. This typical appearance is due to the ability of the material to seal the marginal gap with greyish corrosion products and has to be carefully considered by the clinician during the decisional process. In the right part of the section, a primary lesion can be seen, developing independently from the existence of a restoration on the occlusal side of the tooth

From a clinical point of view, different materials differ in the behavior of their interfaces. Amalgams produce a greyish halo in the tooth tissues due to the depositions of corrosion products of the material (Fig. 3.5). Glass-ionomer cements, resin-modified glass ionomers, and resin-based composites may exhibit visible gaps and/or marginal staining related to the interface colonization, and biofilm overgrowth in the marginal area [6, 71].

The adhesive restorative materials show a significant influence on the microenvironment of the marginal area and play a crucial role in the determination of the composition and characteristics of oral biofilms in different ways [72–76]. As previously mentioned, the chemical and structural characteristics of the material itself, as well as the application technique and the ability of the operator, contribute to this behavior. The latter two variables deeply influence the biological behavior of restorative materials by multiplying the possible types of interfaces that can be obtained. Indeed, the adhesion process is very susceptible to a moist environment. Therefore, controlling humidity in the operating field is fundamental to improve the longevity of adhesive restorations. The best way to achieve this objective is considered to be the use of the dental dam [77]. Nevertheless, in many countries, a relatively low percentage of dental practitioners use dental dams in adhesive procedures regularly.

The other variable that influences the characteristics of the interface is represented by the ability of the operator. The production of a new dental surface anatomy using the restorative material is often far from the natural one, showing gaps and grooves in the marginal area. The roughness of the material and the tissues surrounding the restoration that are prepared for the adhesive process are frequently much higher than that of the original natural tissues [73, 78]. This problem reduces mechanical biofilm removal both by natural mechanisms and by oral hygiene procedures [79]. Finally, restorative materials differ from natural tissues in several surface parameters such as surface free energy and chemical composition.

In fact, clinical studies reported shorter longevity and higher failure rates for direct adhesive restorations in comparison to amalgam, and secondary caries is the main reason for failure. These data are strictly related to the diagnostic problems posed by secondary caries, as previously explained [71].

A very different way to build an interface with dental hard tissues is represented by materials such as glass ionomers and resin-modified glass ionomers that are designed to release compounds, mainly ions. These materials can bond to tooth surfaces by chemical interaction. They are usually composed of alumino-fluorosilicate glass powder that reacts with an aqueous solution of polyacrylic acid [80]. Glass ionomers can release several ions, including fluoride, already starting from the setting reaction. There is a first quick release, the so-called burst effect, where most of the ions are released in the first 2 days after placement. According to several studies, the amount of fluoride released varies from 5 to 155 ppm. After that, there is a relatively long period of slow but constant release that can last up to 3 years. The release has an effect on the surrounding tissues and the interface and has been related to a protective action of the material against secondary caries occurrence. Furthermore, this kind of restorative material can be recharged by reuptake of fluoride ions from the surrounding environment when the concentration of these ions is increased, for instance, after toothbrushing or the application of fluoride-containing mouth rinses [74, 81]. Nevertheless, from a mechanical and aesthetical point of view, glass ionomers are outperformed by composites, and cannot be used in load-bearing restoration. They showed a higher risk of fracture in extended cavities and cannot adequately support the requested performances in highly aesthetic areas.

Resin-modified glass ionomer cements have been introduced to provide the best of both worlds: the resin components add strength and aesthetics, while the material maintains its ion-releasing capabilities. These materials feature a lower release of fluoride than conventional glass ionomers. Their mechanical and aesthetic properties are nonetheless inferior to those of composites.

Polyacid-modified composites (compomers) were another try in this sense. They do not show an initial fluoride-release burst effect [82] as glass ionomers and RMGICs, yet the levels of released fluoride seem to remain much more constant over time, probably due to the characteristics of the composite resin matrix [83]. A drawback of these materials is that fluoride ions hamper the polymerization processes of the resin matrix, producing materials that reach a suboptimal polymerization degree. This phenomenon, affecting the mechanical characteristics, leads to surface degradation, and has been related to increased microbial colonization. In fact, compomers seem to elicit the highest biofilm formation when compared to the surfaces of the previously mentioned materials [84].

Giomers are a new kind of ion-releasing material, consisting of a composite where the filler particles are made of pre-reacted glass ionomers. Similar to compomers, they do not show an initial "burst" effect regarding the release of fluoride, and the amount of fluoride released is considered to be somewhere halfway between conventional glass ionomers and compomers [85]. These materials have shown notable performances in terms of a reduction of biofilm formation in vitro and in situ [86, 87]. Nevertheless, clinical data supporting the protective action against secondary caries by these materials are still controversial [88]. From the point of view of secondary caries occurrence, glass ionomer restorations unexpectedly showed similar performance when compared with composites. In spite of generally very promising in vitro results, it was not possible to demonstrate for this class of restorative materials a clear protection in vivo against secondary caries occurrence [26, 89]. Furthermore, the last generation of glass ionomer materials, featuring improved performances due to the application of nanotechnologies, has not been sufficiently tested clinically quite yet. Nevertheless, several countries have adopted glass ionomers as the standard amalgam alternative as a result of their easy applicability and low placement costs [71].

New generations of restorative materials based on antifouling approaches, antimicrobial activity and remineralization capabilities showing active interaction both with the surrounding tissues and with biofilms to prevent secondary caries occurrence are further discussed in Chaps. 8, 9, and 10.

3.6 Artificial Biofilm-Induced Secondary Caries Models

The occurrence of secondary caries is related to a plurality of factors, and an increasing amount of data suggests that interactions of these materials with the oral ecosystem and in particular with biofilms play a crucial role. As previously described, these interactions can compromise the integrity of a restoration by modifying the structure and the characteristics of materials, mainly at the surface and interface level (See also Chap. 7). In this perspective, in recent years, the research field of biomaterial-biofilm interactions and of the experimental models to study them has raised increasing interest. While the methods to analyze the physical properties and biocompatibility of restorative materials are relatively standardized, as we are going to see in the next chapter, methods to investigate their microbiological behavior are far from reaching similar levels of standardization [90–93]. Studying biofilms is challenging because of their intrinsic characteristics: they are highly heterogeneous and complex. In addition to that, biofilms represent a living community in rapid evolution and are extremely sensitive to even small changes in the surrounding microenvironment (Chap. 1). For these reasons, high intra-sample and sample-to-sample variabilities are most often found in results.

This situation has led to considerable heterogeneity in experimental setups regarding biofilm growth conditions and quantitative determination methods. A wide range of choices are also available for the microorganisms that should be tested in these models, for example, monospecific vs. microcosm models—which produces a severe problem when comparing the outcomes of studies with different experimental designs. Difficulties trying to summarize and to interpret the results obtained by such different methodologies can lead to misleading conclusions.

Another factor to consider is that few data are available in the literature regarding the influence of restorative materials on the biophysical properties, structure, and composition of biofilms in vivo, yet these are also potentially useful predictors of clinical efficacy.

Currently, the research community is discussing if single- or multispecies biofilm models are more appropriate to study recurrent caries. The experimental model to be used should ultimately depend on the specific aim of a study [94]. A typical example involves the use of single-strain biofilms of *S. mutans*. While it is recognized that monospecies biofilms cannot adequately mimic the complexity of in vivo multispecies ones, they are still a useful simplification of a highly cariogenic biofilm that may provide valuable information that requires, however, interpretation within the limitations of the model. In fact, it should be noted that the single cariogenic species can express a surprisingly wide diversity of phenotypes that can bring significant differences in growth characteristics and metabolism such as acidogenicity/aciduricity and tolerance to oxidative stress [94].

Studies on biomaterial-biofilm interactions can take advantage of multispecies models since they allow them to provide a more accurate approximation of clinical conditions. In this way, results can reach a better translational value, since they can reduce the variability that is a peculiar characteristic of in situ or in vivo studies.

Concluding, the importance of in vitro experimental setups is increasing due to the several advantages that they deliver. Current research in this field is gradually progressing towards a better ability to simulate oral conditions in order to provide faster, more economical, and accurate reproductions of the clinical environment.

References

1. Cenci MS, Pereira-Cenci T, Cury JA, Ten Cate JM. Relationship between gap size and dentine secondary caries formation assessed in a microcosm biofilm model. Caries Res. 2009;43(2):97–102.
2. Nedeljkovic I, Van Landuyt KL. Secondary caries. In: Dental composite materials for direct restorations. Cham: Springer; 2018. p. 235–43.
3. Moraschini V, Fai CK, Alto RM, Dos Santos GO. Amalgam and resin composite longevity of posterior restorations: a systematic review and meta-analysis. J Dent. 2015;43(9):1043–50.
4. Nedeljkovic I, Teughels W, De Munck J, Van Meerbeek B, Van Landuyt KL. Is secondary caries with composites a material-based problem? Dent Mater. 2015;31(11):e247–77.
5. Opdam NJ, van de Sande FH, Bronkhorst E, Cenci MS, Bottenberg P, Pallesen U, et al. Longevity of posterior composite restorations: a systematic review and meta-analysis. J Dent Res. 2014;93(10):943–9.
6. Kidd EA. Diagnosis of secondary caries. J Dent Educ. 2001;65(10):997–1000.
7. Martignon S, Pitts NB, Goffin G, Mazevet M, Douglas GVA, Newton JT, et al. Caries Care practice guide: consensus on evidence into practice. Br Dent J. 2019;227(5):353–62.
8. Nedeljkovic I, De Munck J, Vanloy A, Declerck D, Lambrechts P, Peumans M, et al. Secondary caries: prevalence, characteristics, and approach. Clin Oral Investig. 2020;24(2):683–91.
9. Schwendicke F, Kniess J, Paris S, Blunck U. Margin integrity and secondary caries of lined or non-lined composite and glass hybrid restorations after selective excavation in vitro. Oper Dent. 2017;42(2):155–64.

10. Seemann R, Flury S, Pfefferkorn F, Lussi A, Noack MJ. Restorative dentistry and restorative materials over the next 20 years: a Delphi survey. Dent Mater. 2014;30(4):442–8.
11. Brouwer F, Askar H, Paris S, Schwendicke F. Detecting secondary caries lesions: a systematic review and meta-analysis. J Dent Res. 2016;95(2):143–51.
12. Kidd EA, Joyston-Bechal S, Beighton D. Diagnosis of secondary caries: a laboratory study. Br Dent J. 1994;176(4):135–8, 9.
13. Magalhães CS, Freitas ABDAD, Moreira AN, Ferreira EF. Validity of staining and marginal ditching as criteria for diagnosis of secondary caries around occlusal amalgam restorations: an in vitro study. Braz Dent J. 2009;20(4):307–13.
14. Mjör IA, Toffentti F. Secondary caries: a literature review with case reports. Quintessence Int. 2000;31(3):165–79.
15. Boston DW. Initial in vitro evaluation of DIAGNOdent for detecting secondary carious lesions associated with resin composite restorations. Quintessence Int. 2003;34(2):109–16.
16. Elderton RJ. Clinical studies concerning re-restoration of teeth. Adv Dent Res. 1990;4(1):4–9.
17. Nuttall N, Elderton RJ. The nature of restorative dental treatment decisions. Br Dent J. 1983;154(11):363–5.
18. Qvist V. Dental caries. The disease and its clinical management. Oxford: Blackwell; 2008.
19. Mjör IA. The reasons for replacement and the age of failed restorations in general dental practice. Acta Odontol Scand. 1997;55(1):58–63.
20. Opdam NJ, Bronkhorst EM, Roeters JM, Loomans BA. A retrospective clinical study on longevity of posterior composite and amalgam restorations. Dent Mater. 2007;23(1):2–8.
21. Palotie U, Vehkalahti MM. Reasons for replacement of restorations: dentists' perceptions. Acta Odontol Scand. 2012;70(6):485–90.
22. Wang Z, Shen Y, Haapasalo M. Dental materials with antibiofilm properties. Dent Mater. 2014;30(2):e1–e16.
23. Askar H, Tu Y-K, Paris S, Yeh Y-C, Schwendicke F. Risk of caries adjacent to different restoration materials: systematic review of in situ studies. J Dent. 2017;56:1–10.
24. Mo SS, Bao W, Lai GY, Wang J, Li MY. The microfloral analysis of secondary caries biofilm around class I and class II composite and amalgam fillings. BMC Infect Dis. 2010;10:241.
25. Delaviz Y, Finer Y, Santerre JP. Biodegradation of resin composites and adhesives by oral bacteria and saliva: a rationale for new material designs that consider the clinical environment and treatment challenges. Dent Mater. 2014;30(1):16–32.
26. Jokstad A. Secondary caries and microleakage. Dent Mater. 2016;32(1):11–25.
27. Chan DC, Hu W, Chung K-H, Larsen R, Jensen S, Cao D, et al. Reactions: antibacterial and bioactive dental restorative materials: do they really work? Am J Dent. 2018;31(Sp Is B):32B–6B.
28. Ibrahim MS, Garcia IM, Kensara A, Balhaddad A, Collares FM, Williams MA, et al. How we are assessing the developing antibacterial resin-based dental materials? A scoping review. J Dent. 2020;93:103369.
29. Imazato S. Antibacterial properties of resin composites and dentin bonding systems. Dent Mater. 2003;19(6):449–57.
30. Makvandi P, Jamaledin R, Jabbari M, Nikfarjam N, Borzacchiello A. Antibacterial quaternary ammonium compounds in dental materials: a systematic review. Dent Mater. 2018;34(6):851–67.
31. Demarco FF, Collares K, Correa MB, Cenci MS, Moraes RRD, Opdam N. Should my composite restorations last forever? Why are they failing? Braz Oral Res. 2017;31(s1):e56.
32. Featherstone JD. The caries balance: the basis for caries management by risk assessment. Oral Health Prev Dent. 2004;2(Suppl 1):259–64.
33. Loesche WJ. Role of Streptococcus mutans in human dental decay. Microbiol Rev. 1986;50(4):353–80.
34. Kidd E, Joyston-Bechal S, Beighton D. Microbiological validation of assessments of caries activity during cavity preparation. Caries Res. 1993;27(5):402–8.
35. Thomas R, Van Der Mei H, Van Der Veen M, De Soet J, Huysmans MCDNJM. Bacterial composition and red fluorescence of plaque in relation to primary and secondary caries next to composite: an in situ study. Oral Microbiol Immunol. 2008;23(1):7–13.
36. Mejàre I, Malmgren B. Clinical and radiographic appearance of proximal carious lesions at the time of operative treatment in young permanent teeth. Scand J Dent Res. 1986;94(1):19–26.
37. Splieth C, Bernhardt O, Heinrich A, Bernhardt H, Meyer G. Anaerobic microflora under class I and class II composite and amalgam restorations. Quintessence Int. 2003;34(7):497–503.
38. Marsh PD. Dental plaque as a biofilm and a microbial community—implications for health and disease. BMC Oral Health. 2006;6(s1):S14.
39. Gonzalez-Cabezas C, Li Y, Gregory RL, Stookey GK. Distribution of three cariogenic bacteria in secondary carious lesions around amalgam restorations. Caries Res. 1999;33(5):357–65.
40. Beighton D. The complex oral microflora of high-risk individuals and groups and its role in the caries process. Community Dent Oral Epidemiol. 2005;33(4):248–55.
41. Nikawa H, Yamashiro H, Makihira S, Nishimura M, Egusa H, Furukawa M, et al. In vitro cariogenic potential of Candida albicans. Mycoses. 2003;46(11–12):471–8.
42. Klinke T, Kneist S, de Soet JJ, Kuhlisch E, Mauersberger S, Forster A, et al. Acid production by oral strains of Candida albicans and lactobacilli. Caries Res. 2009;43(2):83–91.
43. Hals E, Nernaes A. Histopathology of in vitro caries developing around silver amalgam fillings. Caries Res. 1971;5(1):58–77.

44. Schwendicke F. Contemporary concepts in carious tissue removal: a review. J Esthet Restor Dent. 2017;29(6):403–8.
45. Innes NP, Frencken JE, Schwendicke F. Don't know, can't do, won't change: barriers to moving knowledge to action in managing the carious lesion. J Dent Res. 2016;95(5):485–6.
46. Mertz-Fairhurst EJ, Curtis JW Jr, Ergle JW, Rueggeberg FA, Adair S. Ultraconservative and cariostatic sealed restorations: results at year 10. J Am Dent Assoc. 1998;129(1):55–66.
47. Hals E, Andreassen BH, Bie T. Histopathology of natural caries around silver amalgam fillings. Caries Res. 1974;8(4):343–58.
48. Kidd EA. Microleakage in relation to amalgam and composite restorations. A laboratory study. Br Dent J. 1976;141(10):305–10.
49. Kidd EA, Joyston-Bechal S, Beighton D. Marginal ditching and staining as a predictor of secondary caries around amalgam restorations: a clinical and microbiological study. J Dent Res. 1995;74(5):1206–11.
50. Kuper NK, van de Sande FH, Opdam NJ, Bronkhorst EM, de Soet JJ, Cenci MS, et al. Restoration materials and secondary caries using an in vitro biofilm model. J Dent Res. 2015;94(1):62–8.
51. Thomas R, Ruben J, Ten Bosch J, Fidler V, Huysmans MCDNJM. Approximal secondary caries lesion progression, a 20-week in situ study. Caries Res. 2007;41(5):399–405.
52. Gordan VV, Garvan CW, Richman JS, Fellows JL, Rindal DB, Qvist V, et al. How dentists diagnose and treat defective restorations: evidence from the dental practice-based research network. Oper Dent. 2009;34(6):664–73.
53. Nassar HM, Gonzalez-Cabezas C. Effect of gap geometry on secondary caries wall lesion development. Caries Res. 2011;45(4):346–52.
54. Totiam P, Gonzalez-Cabezas C, Fontana MR, Zero DT. A new in vitro model to study the relationship of gap size and secondary caries. Caries Res. 2007;41(6):467–73.
55. Kidd E, Fejerskov O. What constitutes dental caries? Histopathology of carious enamel and dentin related to the action of cariogenic biofilms. J Dent Res. 2004;83(s1):35–8.
56. Kidd E, O'Hara J. The caries status of occlusal amalgam restorations with marginal defects. J Dent Res. 1990;69(6):1275–7.
57. Pimenta LA, Navarro MF, Consolaro A. Secondary caries around amalgam restorations. J Prosthet Dent. 1995;74(3):219–22.
58. Rezwani-Kaminski T, Kamann W, Gaengler P. Secondary caries susceptibility of teeth with long-term performing composite restorations. J Oral Rehabil. 2002;29(12):1131–8.
59. Hodges DJ, Mangum FI, Ward MT. Relationship between gap width and recurrent dental caries beneath occlusal margins of amalgam restorations. Community Dent Oral Epidemiol. 1995;23(4):200–4.
60. Diercke K, Lussi A, Kersten T, Seemann R. Isolated development of inner (wall) caries like lesions in a bacterial-based in vitro model. Clin Oral Investig. 2009;13(4):439–44.
61. Hayati F, Okada A, Kitasako Y, Tagami J, Matin K. An artificial biofilm induced secondary caries model for in vitro studies. Aust Dent J. 2011;56(1):40–7.
62. Maske TT, van de Sande FH, Arthur RA, Huysmans M, Cenci MS. In vitro biofilm models to study dental caries: a systematic review. Biofouling. 2017;33(8):661–75.
63. Montagner AF, Kuper NK, Opdam NJ, Bronkhorst EM, Cenci MS, Huysmans MC. Wall-lesion development in gaps: the role of the adhesive bonding material. J Dent. 2015;43(8):1007–12.
64. Montagner AF, Opdam NJ, Ruben JL, Bronkhorst EM, Cenci MS, Huysmans MC. Behavior of failed bonded interfaces under in vitro cariogenic challenge. Dent Mater. 2016;32(5):668–75.
65. Ferracane JL. Resin composite—state of the art. Dent Mater. 2011;27(1):29–38.
66. Lynch CD, Frazier KB, McConnell R, Blum I, Wilson NHF. State-of-the-art techniques in operative dentistry: contemporary teaching of posterior composites in UK and Irish dental schools. Br Dent J. 2010;209(3):129.
67. Chin G, Chong J, Kluczewska A, Lau A, Gorjy S, Tennant M. The environmental effects of dental amalgam. Aust Dent J. 2000;45(4):246–9.
68. Hörsted-Bindslev P. Amalgam toxicity—environmental and occupational hazards. J Dent. 2004;32(5):359–65.
69. Meerbeek BV, Yoshihara K, Van Landuyt K, Yoshida Y, Peumans M. From Buonocore's pioneering acid-etch technique to self-adhering restoratives. A status perspective of rapidly advancing dental adhesive technology. J Adhes Dent. 2020;22(1):7–34.
70. Walsh L, Brostek AM. Minimum intervention dentistry principles and objectives. Aust Dent J. 2013;58:3–16.
71. Askar H, Krois J, Gostemeyer G, Bottenberg P, Zero D, Banerjee A, et al. Secondary caries: what is it, and how it can be controlled, detected, and managed? Clin Oral Investig. 2020;24(5):1869–76.
72. Chenicheri S, Usha R, Ramachandran R, Thomas V, Wood A. Insight into oral biofilm: primary, secondary and residual caries and phyto-challenged solutions. Open Dent J. 2017;11:312–33.
73. Cazzaniga G, Ottobelli M, Ionescu AC, Paolone G, Gherlone E, Ferracane JL, et al. In vitro biofilm formation on resin-based composites after different finishing and polishing procedures. J Dent. 2017;67:43–52.
74. Ionescu A, Brambilla E, Hahnel S. Does recharging dental restorative materials with fluoride influence biofilm formation? Dent Mater. 2019;35(10):1450–63.
75. Ionescu AC, Cazzaniga G, Ottobelli M, Ferracane JL, Paolone G, Brambilla E. In vitro biofilm formation on resin-based composites cured under different surface conditions. J Dent. 2018;77:78–86.

76. Ionescu AC, Hahnel S, Konig A, Brambilla E. Resin composite blocks for dental CAD/CAM applications reduce biofilm formation in vitro. Dent Mater. 2020;36(5):603–16.
77. Marshall K. Rubber dam. Br Dent J. 1998;184(5):218–9.
78. Teughels W, Van Assche N, Sliepen I, Quirynen M. Effect of material characteristics and/or surface topography on biofilm development. Clin Oral Implants Res. 2006;17(Suppl 2):68–81.
79. Øilo M, Bakken V. Biofilm and dental biomaterials. Materials. 2015;8(6):2887–900.
80. Neti B, Sayana G, Muddala L, Mantena SR, Yarram A, Harsha GVD. Fluoride releasing restorative materials: a review. Int J Dent Mater. 2020;2(1):19–23.
81. Attar N, Turgut MD. Fluoride release and uptake capacities of fluoride-releasing restorative materials. Oper Dent. 2003;28(4):395–402.
82. Yip H-K, Smales RJ. Fluoride release from a polyacid-modified resin composite and 3 resin-modified glass-ionomer materials. Quintessence Int. 2000;31(4):261–6.
83. Vermeersch G, Leloup G, Vreven J. Fluoride release from glass–ionomer cements, compomers and resin composites. J Oral Rehabil. 2001;28(1):26–32.
84. Brambilla E, Cagetti MG, Gagliani M, Fadini L, Garcia-Godoy F, Strohmenger L. Influence of different adhesive restorative materials on mutans streptococci colonization. Am J Dent. 2005;18(3):173–6.
85. Itota T, Carrick TE, Yoshiyama M, McCabe JF. Fluoride release and recharge in giomer, compomer and resin composite. Dent Mater. 2004;20(9):789–95.
86. Hahnel S, Ionescu AC, Cazzaniga G, Ottobelli M, Brambilla E. Biofilm formation and release of fluoride from dental restorative materials in relation to their surface properties. J Dent. 2017;60:14–24.
87. Hahnel S, Wastl DS, Schneider-Feyrer S, Giessibl FJ, Brambilla E, Cazzaniga G, et al. Streptococcus mutans biofilm formation and release of fluoride from experimental resin-based composites depending on surface treatment and S-PRG filler particle fraction. J Adhes Dent. 2014;16(4):313–21.
88. van de Sande FH, Opdam NJ, Truin GJ, Bronkhorst EM, de Soet JJ, Cenci MS, Huysmans M-C. The influence of different restorative materials on secondary caries development in situ. J Dent. 2014;42(9):1171–7.
89. Randall RC, Wilson NHF. Glass-ionomer restoratives: a systematic review of a secondary caries treatment effect. J Dent Res. 1999;78:628–37.
90. de Souza Costa CA, Hebling J, Scheffel DL, Soares DG, Basso FG, Ribeiro APDJDM. Methods to evaluate and strategies to improve the biocompatibility of dental materials and operative techniques. Dent Mater. 2014;30(7):769–84.
91. Ferracane JL. Models of caries formation around dental composite restorations. J Dent Res. 2017;96(4):364–71.
92. Gomes IB, Meireles A, Gonçalves AL, Goeres DM, Sjollema J, Simões LC, et al. Standardized reactors for the study of medical biofilms: a review of the principles and latest modifications. Crit Rev Biotechnol. 2018;38(5):657–70.
93. Heintze SD, Ilie N, Hickel R, Reis A, Loguercio A, Rousson V. Laboratory mechanical parameters of composite resins and their relation to fractures and wear in clinical trials—a systematic review. Dent Mater. 2017;33(3):e101–e14.
94. Salli KM, Ouwehand AC. The use of in vitro model systems to study dental biofilms associated with caries: a short review. J Oral Microbiol. 2015;7(1):26149.

Bioreactors: How to Study Biofilms In Vitro

4

Andrei Cristian Ionescu and Eugenio Brambilla

Abstract

The interactions taking place between a dental (bio)material, the surrounding tissues of the host, and the biofilm that grows to permanently colonize this microenvironment are amazingly complex when analyzed in detail yet contribute to a crucial factor: the balance between health and disease conditions. From a microbiological point of view, this has a dramatic impact on the longevity of dental treatments. Researchers have long since tried to recreate, even if in parts, this complexity on a bench, both using a reductionistic approach as often performed in research and, more recently, by trying to create models approaching the most realistic behavior. These efforts yielded a wide range of bioreactor systems currently available. We hope that in a future not too far, bioreactor models will be able to reliably reproduce most clinical conditions, dramatically reducing the need for animal and clinical studies. Unfortunately, a universal bioreactor able to mimic any clinical situation still does not exist. Each model comes entwined with its advantages and limitations that must be acknowledged when choosing which model best fits a distinct experimental design. This situation, together with a reduced overall level of standardization, makes the comparison of the obtained results very difficult. This chapter presents an overview of the microbial communities and the bioreactor models that are most significant for studying the microbiological performances of dental materials.

4.1 Introduction: The Need for Modeling Biofilms in the Lab

Teeth and any dental restorative material, including fixed and removable prosthodontic devices, are non-shedding surfaces, unlike the rest of the surfaces of our body that come into contact with the external environment. As explained in Chaps. 1 and 2, this leads to a unique sequence of events that begins with salivary pellicle formation on intraoral surfaces and finally leads to the development of a mature microbial biofilm firmly attached to these substrates. The presence of shear stresses is one of the most critical driving forces that modulate biofilm formation in the oral environment. In fact, it is primarily responsible for microbial growth as a biofilm community instead of planktonic cells, which can be easily washed away. In this sense, analysis of the fluidodynamics at the interface between microorgan-

A. C. Ionescu (✉) · E. Brambilla
Oral Microbiology and Biomaterials Laboratory, Department of Biomedical, Surgical, and Dental Sciences, University of Milan, Milan, Italy
e-mail: andrei.ionescu@unimi.it; eugenio.brambilla@unimi.it

A. C. Ionescu, S. Hahnel (eds.), *Oral Biofilms and Modern Dental Materials*, https://doi.org/10.1007/978-3-030-67388-8_4

isms and hard surfaces is critical to explain many of the fundamental aspects of dental biofilms [1].

In this confined yet highly dynamic microenvironment, microorganisms, surface characteristics of the interface, an array of factors deriving from the host, and external factors such as, most importantly, nutrient intake all contribute to biofilm formation. All of these factors are involved in a biofilm's community balance between health and disease conditions [2]. It is easy to understand that this system has an extreme implicit complexity and is also responsible for the very high inter- and intraindividual variability commonly observed [3, 4]. The design of most in vivo studies dealing with biofilm formation collides with this complexity even if only relatively simple research questions shall be answered. It is also noteworthy that many of the novel materials and technologies that are developed in a struggle to control and modulate microbial colonization and biofilm formation cannot be directly applied in vivo as a result of obvious ethical concerns.

A major part of the hospital-acquired infections is due to biofilm-forming pathogens [5]. Almost all infections of temporary and permanent indwelling devices are characterized by biofilm formation [6]. Many different bacterial species, such as *Pseudomonas aeruginosa*, *Staphylococcus aureus*, and even saprophyte species such as *Candida albicans*, are associated with biofilm infections of indwelling devices that can lead to the chronicization of a disease or to complete failure of the therapy in many different regions of the human body. To reduce the occurrence of such adverse events, the study of biofilms in the medical setting is, therefore, of highest importance. The in vivo approach to study biofilms is still extremely challenging due to the reduced possibility of controlling experimental parameters and, again, to the indispensable ethical concerns that may arise [7]. New strategies are required to simulate the clinical situation in vitro, and several experimental data have been published in the last years on biofilm formation under different conditions and strategies aimed to control their colonization of human tissues [8–10]. Several types of artificial systems, called bioreactors, have been proposed for this issue; basically, they try to mimic the environmental conditions of biofilm development on the surface or inside the human body. The ultimate aim of the bioreactors is to obtain biofilm structures that are functionally and morphologically similar to those found in health or disease conditions, by reproducing most of the conditions found in the human body [2, 11]. Most of the parameters that define these conditions are nowadays reproducible in vitro—for instance, the use of media that simulate the human fluid composition, its flow, the presence of nutrients, the oxygen levels, the adherence and growth substrates, and the temperature (Fig. 4.1). However,

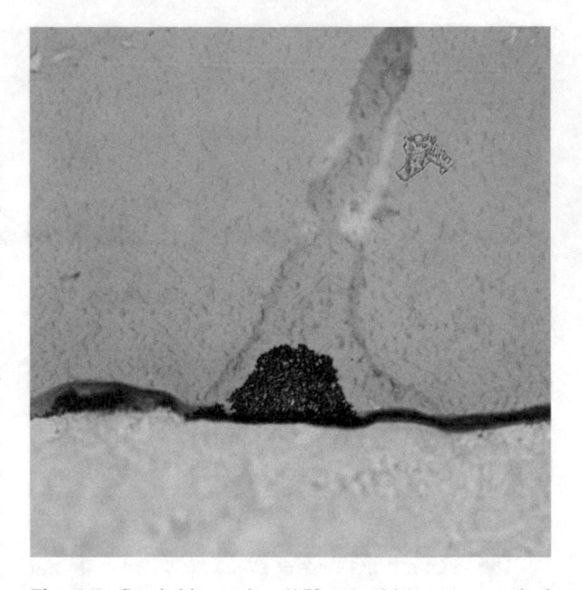

Fig. 4.1 Semi-thin section (150 μm wide) seen at optical microscopy of a *S. mutans* microcolony developing over the surface of the resin component of a dentin-bonding system (1:1 vol BisGMA:TEGDMA resin). The specimen was cultured in a continuous-flow bioreactor (MDFR) for 96 h. Cells and extracellular matrix are colored in violet. It can be seen that, immediately after adhering to the surface, bacteria start replicating forming a monolayer and producing the extracellular matrix in which they are embedded and that protects them. After that, biofilm formation takes place with the development of microcolonies and the production of an excess of extracellular matrix that forms a "tail," here stained in light violet. The latter originates from the microcolony and is situated in an upward position due to the lack of hydrodynamic shear during specimen processing. Under flow conditions, the tail is oriented downstream and can be detached by high shear stresses or by the "decision" of the bacteria themselves through quorum sensing to depolymerize the extracellular matrix to be able to go and colonize other surfaces downstream. The necessity of replicating such behavior in vitro is paramount to approach the clinical behavior of the studied biofilms. (Specimen preparation and observation courtesy of Dr. Vincenzo Conte and Prof. Patrizia Procacci, University of Milan, Italy)

some parameters and conditions are still considerably challenging to reproduce; this includes, for example, the host immune response. The latter has a crucial influence on the growth and structure of biofilms, yet this interaction is still not possible to be reproduced in vitro.

Despite that, the bioreactors allow for testing of a relatively large number of specimens under very defined conditions. The variability associated with the environmental conditions is thus significantly reduced, and the experimental parameters that are studied can be reliably controlled [7, 11, 12].

In the oral environment, biofilm development is a commonly occurring event. Researchers have attempted for years to find a way to disrupt and prevent biofilm formation, with generally poor results. The current trend is, on the contrary, to modulate the behavior of the oral biofilm in order to favor the growth of nonpathogenic species selectively and to reduce the development and metabolism of pathogenic ones. This ecological perspective on biofilm studies has a deep impact on in vitro modeling since the whole complexity of the multispecies oral microflora has to be consistently reproduced and maintained for the desired experimental duration [13, 14]. This approach needs to be matched with sophisticated methodologies that are capable of assessing the prevalence of the different components of the microflora. Such technologies have only been available for a few years and add to the complexity of these studies [15, 16].

Bioreactors, when coupled with specific instruments for measuring biofilm characteristics, can be used as tools to "sense" the behavior of the microenvironment in a more subtle way than many modern instruments [10, 17, 18]. These setups can use growth conditions and parameters that are, on purpose, far from clinical situations. In this way, minimal amounts of drug release can be detected, as well as material surface modifications, and even accelerated aging of the exposed interfaces can be simulated (Fig. 4.2). For instance, considering caries research, a recent study highlighted that the effect of fluoride on *S. mutans* biofilm formation is dependent on the bacterial strain that is employed [19].

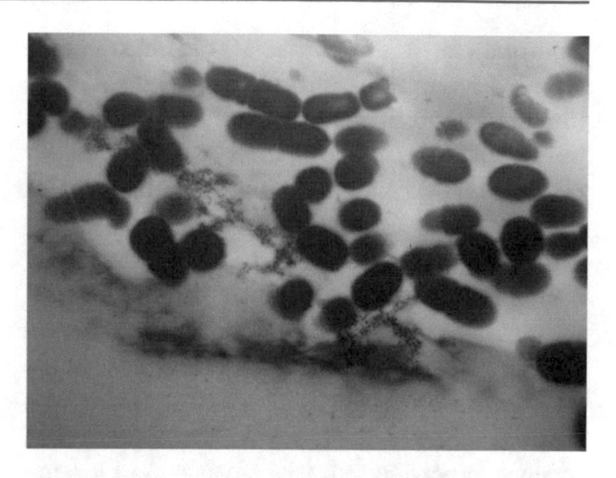

Fig. 4.2 Semi-thin section of the previous specimen observed using transmission microscopy after 96 h of biofilm formation. A nutrient medium (undefined mucin medium) highly enriched in sucrose (5 wt.%) continuously fed through the bioreactor inlet causes extra production of acidic catabolites, extracellular matrix, and, possibly, esterases by *S. mutans* cells. This situation is far from clinical situations where biofilms are not composed by a single species, pH close to the surface does not reach such low values for such extended time, and a human being is not continuously fed with high amounts of simple carbohydrates as its only nutritional intake. Nevertheless, these extremized conditions of "accelerated microbiological aging" show the initial degradation of the resin surface that is expressed as an initial staining of subsurface layer with the hydrophilic electron-dense dyes, lead citrate and uranyl acetate. This type of study can provide further insight, for instance, on the microbiological corrosion and deterioration of dental materials, and the microbiological reasons for failure of an adhesive interface between a resin-based composite restoration and natural tooth tissues. (Specimen preparation and observation courtesy of Dr. Vincenzo Conte and Prof. Patrizia Procacci)

4.2 The Choice of the Microbial Community

A broad range of bioreactor devices and systems is currently available for the investigation of oral biofilms. Nevertheless, strategic choices must be performed before selecting a specific device. Oral biofilms are complex communities in which hundreds of species coexist in the same ecological niche, expressing synergistic or antagonistic behavior among them, while, at the same time, establishing a symbiotic relationship with the host [20]. The selection of a specific inoculum depends on the individual requirements of the study or the research question. The microbiological model that most closely simulates this microenvironment is the artificial oral microcosm [21–23]. Microcosms are microbial communities that are grown in vitro to replicate as closely as possible the behavior of their in vivo counterparts. They have a microbiological composition similar to that of the oral environment they are replicating, and this is usually obtained by using biofilms that are sampled from the oral environment. Also, particular care is necessary to ensure that the experimental setup precisely reproduces the physicochemical conditions as well as the nutrient composition. Experiments performed using microcosms can take advantage of a setup that is quite similar to the oral environment, which enables the evaluation of the dynamic performance of the microbial community and ensures control over the experimental parameters that are studied. Dental plaque microcosms were used to provide a better knowledge of the microbial ecology and physiology of dental microbial ecosystems [11, 24–26] (Fig. 4.3).

Fig. 4.3 An example of the complexity of the interactions between biofilms and dental materials' surfaces. Confocal laser scanning microscopy was used to obtain a 3D reconstruction of an artificial oral microcosm grown in a bioreactor (MDFR) over a non-buffering surface of a conventional resin-based composite material. LIVE-DEAD stain used Syto-9 and propidium iodide to stain viable cells in green and dead cells in red, respectively. A thin layer of dead cells can be identified close to the surface, while the more external layers are all made of viable microbial cells. No antimicrobial compounds were used on this materials' surface, yet the combination of reduced amount of nutrients and decreased clearance of acidic catabolites (that are not buffered by demineralization as happens on natural surfaces) makes the microenvironmental conditions close to the surface very hostile. From this point of view, the presence of a "tamper" layer of dead cells may be highly detrimental to the equilibrium between health and disease conditions, since, being dead indeed, it does not react. It may thus greatly prolong the contact of acidic catabolites and degradation compounds such as esterases with the surface, accelerating the deterioration of the material and secondary caries onset

There are, however, limitations related to the use of microcosms. The microbial communities have huge variability in composition due to site- and subject-specific heterogeneity of the inocula. This circumstance produces variable results when comparing results for different experimental runs and raises difficulties regarding the comparison of the results obtained by different workgroups. Specific microbial species whose presence might be essential to the experiment may not be present in the inoculum, while the presence of undesirable species may unpredictably influence the outcomes. It has to be noted, however, that microbial communities have an intrinsic capacity of adaptation that strictly depends on the microenvironmental conditions. Therefore, the latter may lead to communities expressing similar phenotypical behavior, even if the starting inocula are different. An example can be seen in the massive selective pressure that the presence of sucrose exerts on microbial communities, shifting their composition towards the prevalence of acidogenic species. The composition of microbial communities, however, cannot be easily controlled to comply with the experimental objectives, and this type of inoculum is also the most difficult to standardize [27, 28].

A simplification criterium can be applied to reproduce this complex microenvironment only in parts in order to comply with specific research questions. To do that, researchers are modeling biofilms made of single species, or defined consortia made of few species growing together. While these approaches may seem outdated nowadays, they still provide significant advantages over the more complex microcosm models. A reductionistic approach can be efficiently used to control the influence of single parameters and for screening purposes—for instance, when the influence of a wide array of active principles or adherence substrates has to be tested. An example can be the initial testing of an array of active principles that are intended to be incorporated into a dental material. Several compositions and concentrations have to be tested in the most efficient and less time-consuming way to select the most promising ones.

Defined consortia of few species can provide a simplified simulation of ecological phenomena that are relatively easy to study due to the known parameters such as the initial and final proportion of the different species. The use of defined consortia is based on the evidence that many biofilm-generated diseases are a result of the combined activity of a group of microbial species in which each member is only weakly virulent. Each species can play a specific role or function, allowing the consortium to persist and express pathogenicity [2, 29]. Recent findings have proposed the concept of low-abundance species, due to which few distinct pathogens are mainly responsible for the virulence of the whole community [30, 31].

Experiments performed using defined consortia and monospecies usually achieve a higher degree of reproducibility compared to microcosm-based biofilms, theoretically allowing for better comparison between experimental runs and among research groups. Many different defined consortia have been developed; nevertheless, literature data show that each research group developed consortia showing different compositions from one another. Thus, the lack of well-defined standard procedures makes comparisons among research groups somehow tricky. One of the first and most used defined consortium models is the "Marsh Consortium" [32]. It is composed of ten microbial species that were chosen to represent the main physiological and ecological groups within the oral cavity. The model has shown excellent stability over time and allows for relatively simple sampling. Many similar approaches have been developed over time [33–35].

The highest degree of simplification can be achieved when using monospecies biofilms. A trade-off in the simplicity of the microbiological approach can bring advantages in terms of standardization and experimental control, making experimental design and interpretation of the results more straightforward [36]. A single-species biofilm is definitely less complex but can provide outcomes that can be useful to develop assays or analytical techniques. It can also be applied when approaches to treat biofilms are targeted towards eradication rather than modulating. For instance, this is the case when surface modifications of a material are performed with the aim of preventing microbial adherence and biofilm formation. One possible strategy is to engineer a material both regarding its surface and its releasing capabilities based on the response to the "pioneer" bacteria, making the surfaces hostile for the first colonizers, thus hoping to prevent the development of a fully mature biofilm. Furthermore, monospecific biofilms are better indicated when specific physiological aspects of the biofilm are to be studied by evaluating the response of the test inoculum to defined experimental conditions. One of the most used monospecies models in caries research is based on *Streptococcus mutans* [37–39]. This species has been identified as one of the main agents associated with dental caries [40]. Its ability to produce large amounts of extracellular matrix makes it able to adhere stably and quickly colonize a wide variety of surfaces, including natural and artificial ones. Moreover, its acidogenicity confers to its biofilm the pathogenic characteristics that are essential in caries research [41–43]. The major limitation of monospecies biofilm models is that they do not exist in the mouth. In fact, *S. mutans* can be a minority species even in persons with active caries [44, 45] and is currently regarded as a marker of caries risk rather than the responsible agent for dental caries.

4.3 Types of Bioreactors

Many bioreactor models are available nowadays. The main difference among them can be drawn between static and dynamic bioreactors. Static bioreactors can still be used to study adhesion and early colonization steps. In the oral environment, biofilm formation is subjected to hydrodynamic stresses. Therefore, the subsequent stages of this process (i.e., biofilm formation) have to be studied with the use of more complex systems that are able to replicate these conditions. Furthermore, the mouth is a very complex environment that can be regarded as an open system, where there is an intermittent inlet of nutrients and a salivary flow that provides clearance and discards catabolites that are produced by microbial metabolism. Bioreactor systems able to reproduce these conditions have evolved into very sophisticated devices that can recently include microfluidic technologies. The difficulty in performing experiments using these devices is proportional to the complexity of such systems. For example, salivary flow and shear forces must be reduced to a minimum during the night, when there is no inlet of nutrients for an extended amount of time. This situation highlights the need for those systems to show a flexible operational envelope. The main types of bioreactors and their application will be shortly discussed, starting from basic designs to the ones with increased complexity.

Fig. 4.4 Agar plate with selective medium for lactobacilli. A serial dilution shows colonies of *Lactobacillus rhamnosus* SD11, a probiotic strain whose presence in oral biofilms is considered caries protective

4.3.1 Static Bioreactor Models

Agar plates are the simplest static model conceived and were used for long to mimic, to some extent, biofilm growth conditions at an air/substrate interface. The finite availability of nutrients poses an intrinsic limit to the biofilm development and to the incubation time. The possibility of this model to evaluate the susceptibility towards different antimicrobial active principles was demonstrated [46, 47]. The availability of nutrients embedded into the substrate makes biofilms developed over agar plate surfaces very different from those growing on hard surfaces, limiting its value when the purpose is to study the interaction of biofilms with the surfaces of dental materials (Fig. 4.4). This situation is more similar to the one occurring when biofilms colonize and infect soft tissues [48, 49]. The agar disk diffusion method for antibacterial compound testing is based on this kind of simple bioreactor model. Nevertheless, the results of this model were not proven to feature a good correlation with in vivo data when considering biofilms developed on the surfaces of indwelling devices [12]. The growth conditions that are reproduced by this model do not show satisfactory similarity with the in vivo clinical situation. An evolution of this model was the colony biofilm method, where biofilm formation was obtained on a semipermeable membrane placed on an agar plate. The usefulness of this model also resides in its use as a preliminary antimicrobial test [7, 50].

A static model that allows a better simulation of microbial adherence and early colonization on hard surfaces is the microtiter plate. This is a simple yet effective closed system that is designed to test a broad array of specimens while keeping control of the growth conditions. A typical assay evaluates the time-dependent adherence to the wells' substrate, which is usually made of polystyrene, polypropylene, or polycarbonate [51,

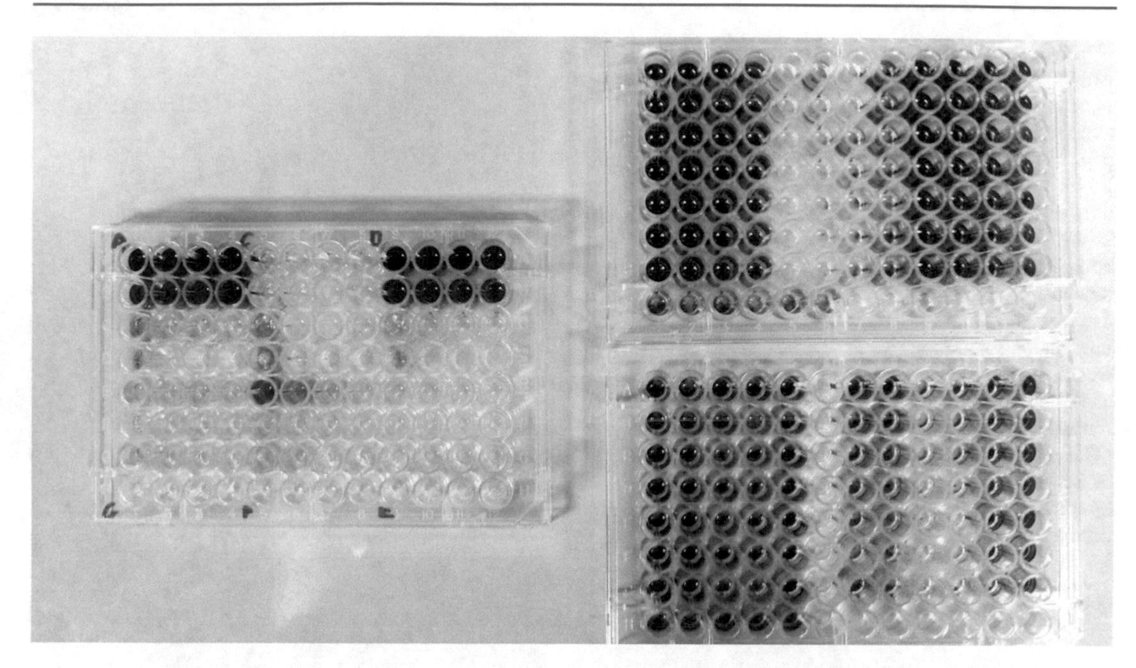

Fig. 4.5 A 96-well microtiter plate test to assess the antimicrobial activity of a library of natural compounds and its derivatives against *S. mutans* biofilms. The adaptability of the system is evident, where multiple replicates can be obtained for each test and parameters such as dilution (for instance, determining the minimal concentration achieving biofilm eradication, MBC), contact time, and activity on different microbial strains can be conveniently studied in a single experimental run. All kind of colorimetric tests can be easily applied and standardized. Here, MTT-based assay, shown on the two plates on the right, is based on the reduction by viable and metabolically active cells of yellow MTT tetrazolium to purple formazan [23]

52]. Furthermore, the substrate can be γ-irradiated to change its surface properties (increase in surface free energy) and better foster cell adherence (tissue culture-treated surfaces). This system can perform preliminary antimicrobial screening tests on a library of compounds (Fig. 4.5). Both the prevention of biofilm formation and the removal potential of antimicrobial compounds can be assessed by the addition of scalar concentrations of test compounds after inoculation or after "mature" biofilms are developed [11, 12]. Care must be taken, however, not to test just a layer of bacteria that is deposited on the bottom of the wells instead of a biofilm. To avoid that, plates must be gently washed at least a couple of times with a buffered isotonic solution to remove non-adhered cells. As such, the microplate model can be coupled with all sorts of high-throughput end-point biochemical quantitative assays, including the evaluation of viable biomass, extracellular matrix, and acid production. Optical measurements using transparent flat-bottomed plates can be performed in real time to plot the growth curves in a nondestructive way [53]. More recently, molecular bioassays can also be performed, for instance, to screen large numbers of strains for specific characteristics [35]. This model is quite a right choice for preliminary testing of dental materials since material samples can be fabricated in a relatively simple way to be press-fitted on the bottom of the plates, or be made with a smaller diameter to allow the collection of the specimen together with the overlying biofilm.

An evolution of the static plate model was developed and patented by the Biofilm Engineering Research Group of Calgary University [54], which is why it was marketed as the Calgary Biofilm Device, now available under the new appellation "MBEC Assay®'s Biofilm Inoculator" that stands for the determination of the minimum biofilm eradication concentration. The static microplate model was modified by adding pegs to the plate lid in correspondence to

Fig. 4.6 A standard MBEC assay 96-well plate is displayed where hydroxyapatite-coated pegs are attached to the lid of the plate and are used as substrate for biofilm formation. It is apparent that biofilms developed on the pegs can be transferred to new plates containing fresh culture medium, or any reagents, by just repositioning the lid. (The picture is courtesy of Dr. Amin Omar, chief operating officer at Innovotech Inc.)

each well. The pegs are used as the substratum for biofilm formation, allowing high-throughput experiments in a simple way (Fig. 4.6). The culture medium can be easily exchanged by transferring the lid to another plate. This constitutes an advantage of this model over the static plate that allows extending the total incubation time well over 24–48 h. In the same way, screening of active principles can be performed without difficulty, and the MBEC can be obtained. Several versions of this device have been proposed by different research groups, with a broad spectrum of substrata, inocula, and growth media. As an example, saliva-coated hydroxyapatite disks were used as a substratum for antimicrobial studies using defined consortia [24, 55]. In this case, the specimens of a dental material to be tested are hanged from the lid and immersed into the culture broth, allowing biofilm to form on their surfaces.

The main drawback of these devices is related to their design, which includes a closed environment with a finite source of nutrients and in which catabolites and eluted compounds become more and more concentrated with time. This situation does not commonly occur in the oral environment. Under these conditions, swift microbial growth occurs in the first moments, followed by a stationary phase. This limitation can nevertheless make this model ideal for measuring the amount of active principles leaking out of the material and concentrating on the supernatant broth, or their activity on the overgrowing biofilms. Furthermore, hydrodynamic stress that is paramount to the development and structure of oral biofilms is absent, which is another drawback. It is clear that the growth of biofilms closely mimicking in vivo conditions requires systems such as intermittent- or continuous-flow devices, where the flow provides nutrients and, at the same time, allows washout of catabolites and eluted compounds. A modification of this model to partially overcome its drawbacks consists of merely inserting the plates into an orbital shaker to provide shear stress. This transforms the model into a straightforward dynamic one that, notwithstanding its still huge limitations such as the presence of a finite amount of nutrients and the radial inhomogeneity of shear stress across the well, allows to provide many of the conditions offered by much more complex dynamic models.

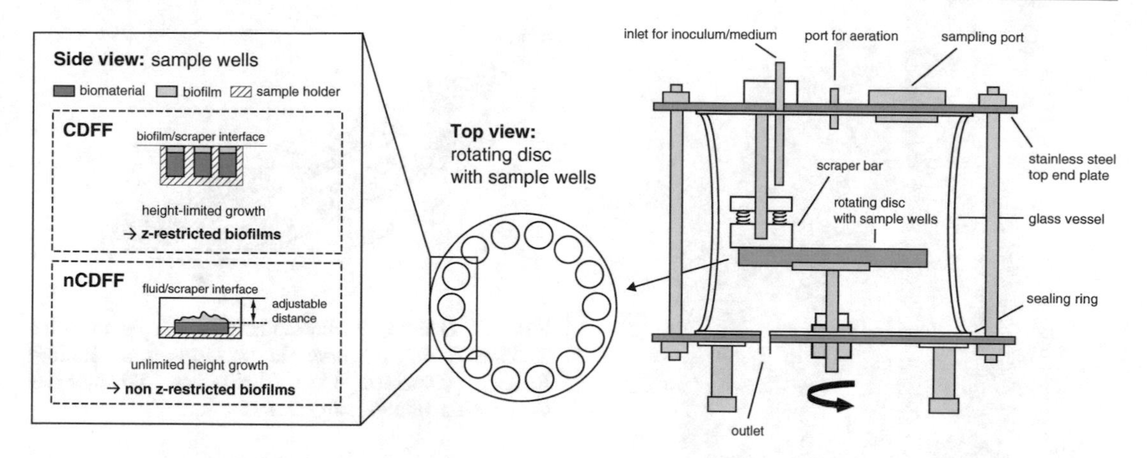

Fig. 4.7 Differences between a standard CDFF and the nCDFF model. (Available from Lüdecke C, Jandt CD, Siegismund D, Kujau MD, Zang E, Rettenmayr M, Bossert J, Roth M. Reproducible Biofilm Cultivation of Chemostat-Grown Escherichia coli and Investigation of Bacterial Adhesion on Biomaterials Using a Non-Constant-Depth Film Fermenter. https://doi.org/10.1371/journal.pone.0084837)

4.3.2 Dynamic Bioreactor Models

A better approximation of the oral environment can only be achieved in vitro by taking into account and replicating the environmental characteristics that influence the growth of oral biofilms. Two main aspects deeply influence oral biofilm development, namely the presence of different interfaces (air/liquid, liquid/substratum) and the hydrodynamic stresses induced by the flow of saliva and nutrients over the substratum surfaces. These aspects determine the transport rate of oxygen, nutrients, active compounds, and catabolites in and out of the biofilm structures. The flow is the primary source of hydrodynamic stress, which is an influential driving factor for the morphology and structure of biofilms. Therefore, it is essential for a bioreactor system to reproduce these conditions in order to develop a biofilm closely resembling in vivo ones.

The research group led by Dr. Philip D. Marsh made a first step approaching the complexity of oral environmental conditions. They developed a continuous culture of oral bacteria in planktonic state and, while the bioreactor was running, they realized that biofilm developed on the vessel walls, possibly simulating dental plaque formation [32]. The research group refined their model by introducing removable hydroxyapatite specimens as growth substratum that were suspended inside the vessel [56]. Furthermore, sucrose addition was performed to select a cariogenic environment.

Another relatively simple approach to model oral biofilm formation was introduced by the constant-depth film fermentor (CDFF) [57, 58]. It consists of a glass vessel with a stainless steel top and bottom plates, containing ports for sampling and inlet/outlet system for nutrients. A high number of specimens can be simultaneously tested (15 PTFE pans allowing 5 specimens each), and the specimens are fitted into the bottom plate. The latter rotates under a scraper blade that helps in diffusing the nutrient medium over the surface of the plate and regulates biofilm depth. The system can be stopped, and sampling pans can be removed aseptically, allowing to study incubation time as a parameter on the same experimental run.

This system was one of the first high-throughput bioreactor devices that allow virtually any substrate to be tested for biofilm formation, providing a suitable platform for the study of the microbiological behavior of dental materials. Great attention was paid afterward in the design of bioreactors to ensure the easiness of testing for different materials. A variant of this model, called nCDFF (nonconstant depth film fermentor), included the possibility to form biofilms without thickness constraint (Fig. 4.7).

Fig. 4.8 Schematic representation of the rotating disk reactor. (Obtained from BioSurface Technologies Corporation. http://biofilms.biz/)

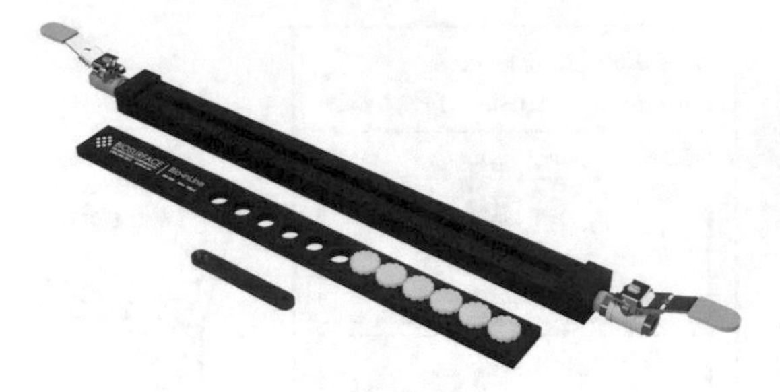

Fig. 4.9 One of the customizations of the modified Robbins device, resulting in the Bio-inLine Biofilm Reactor. (Obtained from BioSurface Technologies Corporation. http://biofilms.biz/)

A variation of this model sharing a similar concept is the rotating disk reactor, RDR (Fig. 4.8). The bioreactor includes a vessel that allows an inlet and outlet of nutrient broth with the presence of a constant amount inside the vessel. At the bottom of the vessel, a magnetic rotor is used as a specimen holder (up to 18 coupons). The hydrodynamic stress generated by this device is easily controlled by adjusting the speed of the rotor. The reactor design was studied to ensure easiness of operation, also including sterilization procedures. The system is flexible, being adaptable to several different studies, ranging from the study of the biofilm exopolysaccharide matrix formation to the rheology of oral biofilms [59, 60]. The operational envelope of this bioreactor was extensively studied, and it was registered as a standard test method for the evaluation of biofilms (ASTM E2196-02). The main advantage of the system includes its simplicity and easiness of use, especially when the hydrodynamic stress parameter is analyzed. The relatively low number of specimens that can be tested at the same time is, however, a limitation.

The Robbins device is a flow-through system also used in medical biofilm studies [61, 62]. It consists of a plastic or metal tube into which specimen-containing coupons can be inserted, becoming part of the tube wall (Fig. 4.9). This system provides similar advantages to the CDFF in terms of high-throughput testing of different substrata and the possibility to aseptically remove every single coupon. Similar to in vivo biofilms, the structure, thickness, and morphology of the biofilm growing on the coupons are influenced by the hydrodynamic parameters of the flow rather than by the scraping activity of a blade, or the velocity of a specimen-holding rotor.

The drip-flow bioreactor was conceived by the Center for Biofilm Engineering of Montana State University [7, 63]. It consists of several parallel independent flow cells that have the dimensions of a microscopy slide. Each flow cell has a lid that can be separately unscrewed to collect the specimens. The name of the reactor is due to the nutrient inlet that drips over the surface of the specimen directly, preventing backward contamination of the tubing. The bioreactor is operated at a 10° inclination so that gravity provides continuous flow with hydrodynamic stress over the specimens' surfaces. The system is therefore used for simulating biofilm formation at the air/liquid interface under relatively low shear stress conditions and allows to study biofilm formation on

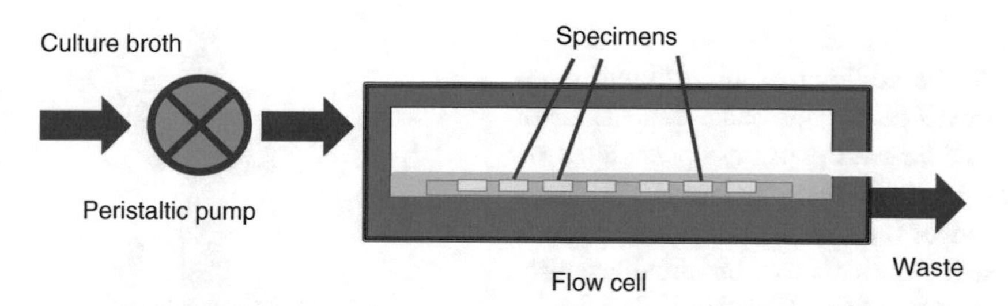

Fig. 4.10 Diagram of the modified drip-flow bioreactor. Teflon holder allows for multiple specimens made of any (bio)material to be immersed into the flowing medium just under the air/liquid interface. The flow cell represents an open circuit, where spent medium is discarded

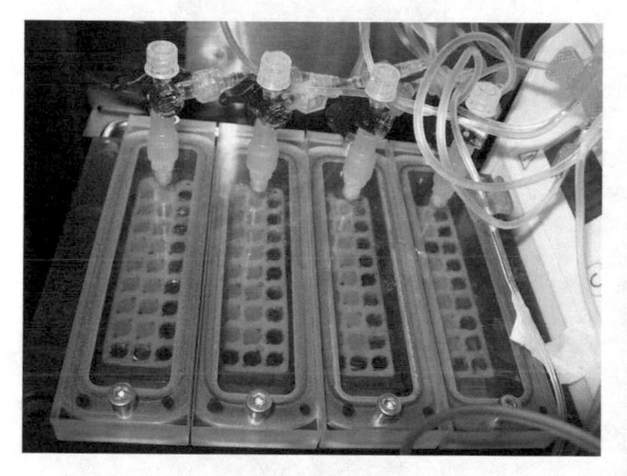

Fig. 4.11 An operating modified drip-flow bioreactor [65]. All tubing are connected through disposable Luer lock and valves, thus ensuring easy and low-cost modifications, such as additional inlets for sucrose pulsing or antimicrobial solution testing

the surface of any material. This system has also been registered as a standard test method for the evaluation of biofilms (ASTM E2647-13). The possibility to easily place or remove specimens allowed the system to be used in several studies that tested the antimicrobial efficacy of oral hygiene products, such as toothpastes or mouthwashes [64–66]. The system still includes some limitations, for instance the low number of specimens that can be tested, the difficulty of temperature control, and the need for complicated, multichannel pumps to operate the flow in parallel to the flow cells reliably.

Brambilla et al. proposed a modification of the drip-flow bioreactor, overcoming some of the limitations of the model [66]. The reactor was operated in a horizontal position, and a low dam was included in the design downstream of the specimen trays to maintain the specimen surfaces immersed in the flowing medium (Figs. 4.10 and 4.11). As with many bioreactors, the system is designed to be entirely placed inside an incubator for optimal temperature control (Fig. 4.12). From the point of view of flow characteristics, this sys-

Fig. 4.12 Placing the whole bioreactor, including the distribution pumps and the main vessels containing the sterile medium, inside an incubator provides optimal temperature control over other more simplistic solutions such as a thermostatic bath or table that often do not allow for homogeneous heat distribution to the whole system [17]

tem shows many similarities with the Robbins device, with the addition of an air/liquid interface. Specimens having the exact dimensions of the bottom of 96-well plates are press-fitted on customized polytetrafluoroethylene (PTFE) trays at the bottom of the flow cells. Up to 27 specimens in each flow cell can be simultaneously tested, allowing this bioreactor to be a high-throughput, very adaptable system for the testing of dental materials [17, 23, 65].

The Center for Disease Control (CDC) developed its own biofilm reactor [22, 67]. It is made of a cylindrical vessel in which eight specimen-containing rods (three specimens per rod) are suspended from the lid. Similar to the rotating disk reactor, an inlet and outlet provide a flow of nutrients and a constant volume is maintained inside the vessel in which specimens are immersed. A magnetic stirrer at the bottom of the vessel can indirectly provide a wide range of hydrodynamic stress by agitating the nutrient broth. Specimens can be assessed at different time points by aseptically removing the rods. This bioreactor was used to provide two standard methods (ASTM E2562-12 and ASTM E2871-13) for biofilm development and test of antimicrobial compounds under high hydrodynamic stress and continuous flow (Fig. 4.13). The system was not initially developed for the study of medical and oral biofilms; therefore several modifications of the system were performed, mainly regarding the growth medium and the control of the temperature and the hydrodynamic flow conditions. Several authors used this system for the development of oral biofilms. Rudney et al. [22] were able to develop oral microcosm biofilms using this model, while Li et al. [68] used the system to study the effect of sucrose pulsing on the biofilm development over the surfaces of dental restorative materials. The main limitation of the system is related to the low amount of specimens that can be tested at the same time.

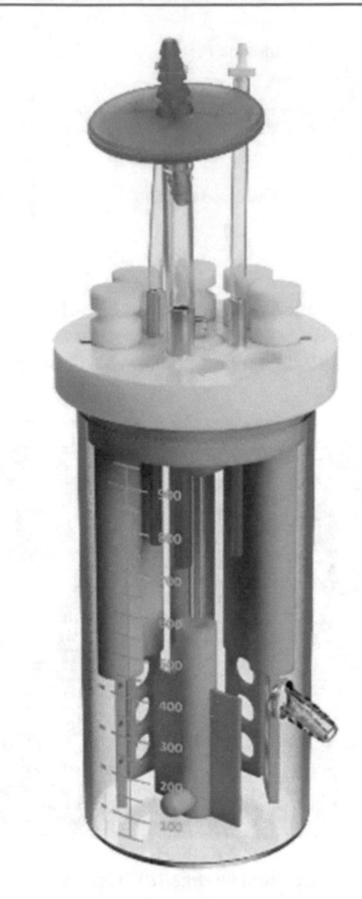

Fig. 4.13 Schematic representation of the CDC Biofilm Reactor. (Obtained from BioSurface Technologies Corporation. http://biofilms.biz/)

4.3.3 Microfluidic Bioreactor Models

More recently, bioreactor systems were developed using microfluidic techniques allowing them to overcome some limitations, such as the relatively large volume of nutrients and biomass that are usually required by the previously described bioreactors. These techniques make it possible to reduce the dimensions of the test environment for better spatial and temporal control of biofilm community formation. Indeed, microfluidic bioreactor systems are small enough to approach the microscopic dimension range. For this reason, they can be efficiently used to study cell interactions during the very first steps in biofilm formation and with the adherence substrate. In the latter case, high interest is due to the study of nanopatterned materials.

Microfluidic devices are built to reproduce the physical effects occurring at the micron scale, including an increase in the surface-to-volume ratio. As a consequence, physical parameters

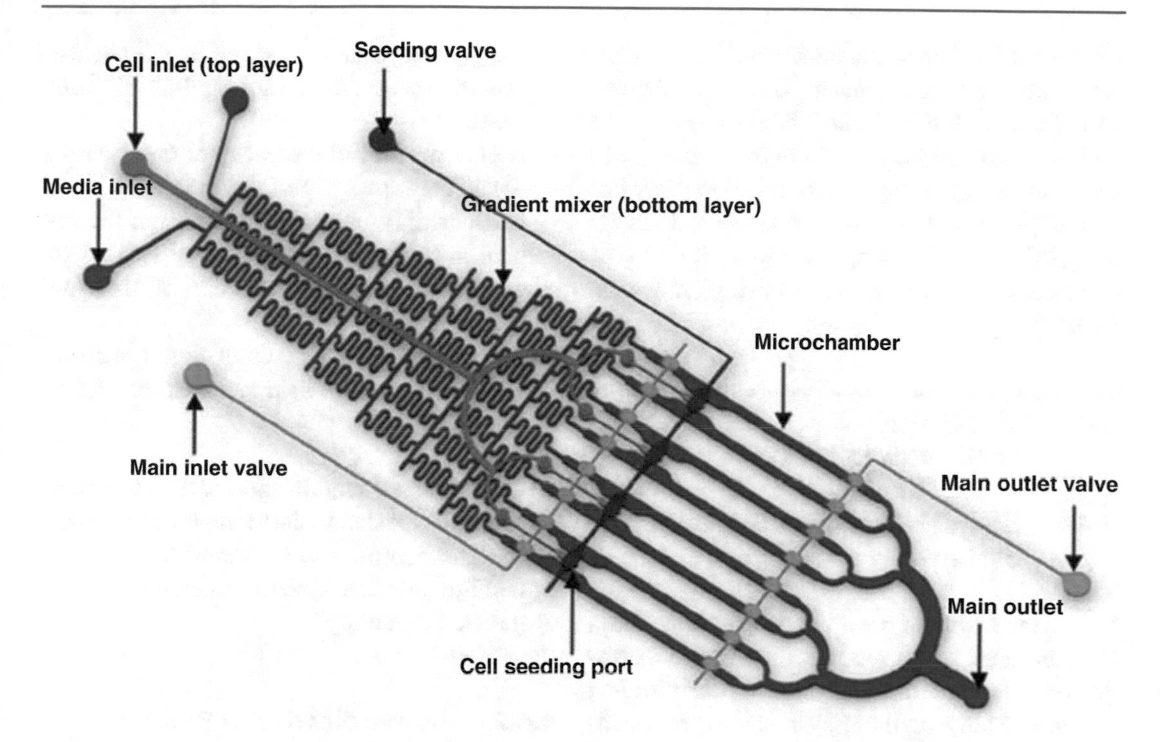

Fig. 4.14 Diagram of the microfluidic device by Jeongyun Kim et al. and its evident complexity. The device consists of a glass coverslip and two PDMS layers—a bottom layer with a diffusive mixer and eight microchambers and a top layer with the pneumatic elements for opening and closing microvalves that separate the diffusive mixer and bacterial seeding ports from the microchambers. The top layer also contains a bacterial seeding port for introducing bacteria into the microchambers

such as capillary forces, fluidic resistance, and surface tension become fundamental in controlling these effects. In particular, laminar flow conditions can be obtained to reach better control. In these conditions, the effect of diffusion becomes predominant over other effects such as turbulence, convection, and gravitational forces. The exchanges of nutrients and catabolites, and, more generally, the energy transfer between a biofilm and the surrounding flow, can be more efficiently controlled and investigated [12, 69, 70]. A very high number of replicates can also be provided for high-throughput analyses. However, the miniaturization of the devices dramatically increases their complexity, which, in turn, increases the difficulty of operating such systems and their inherent costs. In fact, the real microfluidic dynamics of biofilms are very poorly known as it is a relatively new research field. Therefore, no approach is currently able to reproduce real microfluidic conditions and standardization of such systems appears problematic. These systems have been mainly developed for the study of cell cultures, and, then, they were adapted for biofilm development as well. They are often built to provide an answer to a defined research question based on a reductionistic approach rather than to recreate the whole complexity of the clinical situation.

A microfluidic device was developed by Groisman et al. [71] to produce biofilms inside chemostat microchambers, where better control of the microenvironment could be achieved. Kim et al. [72] developed a microfluidic bioreactor based on a two-layer flow cell. The device was built to study the effect of a gradient in the concentration of an active principle or signaling molecule. A total of eight microfluidic flow cells were used to simultaneously expose developed biofilms to different concentrations using a gradient generator based on diffusive mixing (Fig. 4.14). Another device was conceived by Benoit et al. [70] to develop a high number of independent biofilm communities at the same time under a continuous flow using the format of a 96-well plate. This device can be used as a high-throughput system for biofilm screening, and its compatibility with plate readers allows very fast and adaptable biofilm assays. Busscher and van der Mei [73] provided a comprehensive review of flow displacement systems for studying microbial adhesion.

4.4 The Quest for Standardization

4.4.1 Standardization of Bioreactor Systems

It is clear from the previous descriptions that a high number of bioreactor systems are nowadays in use for the analysis of oral biofilms in vitro. Many of these systems are not standardized or have been standardized for different environments rather than the oral environment. A considerable limitation of oral biofilm models has been that, because of their complexity, dynamicity, and adaptation capability, they are difficult to standardize or characterize. Indeed, the validation of a system is much easier than its standardization. The proof of concept and validation of a bioreactor system imply that it works predictably; that is, it is capable of reproducing the desired microenvironment. Also, the repeatability of the results obtained under defined working conditions is ensured. Standardization comes with a higher level of complexity that includes the isolation and investigation of all the possible parameters that may influence the working conditions of the system. The behavior of the system under these conditions (operational envelope) has to be known to control and reduce the sources of variability. A standard method has to comply with all of the following concepts [7]:

- Repeatability (different runs of the bioreactor must produce comparable results)
- Reproducibility (different laboratories using the same system must produce comparable results)
- Ruggedness (minor changes in the standard operating procedure do not significantly affect the results)
- Responsiveness (the capacity of the system to obtain the expected performances)
- Reasonability (any operator can run the system, given specific instructions, without the need for a too high amount of time and consumables)
- Relevance (the outcomes of that system are within the research field to which that system is applied)

Of course, any modification of the operational envelope of a standardized bioreactor system implies that additional studies must be performed to confirm that the system maintains standard operational capability.

4.4.2 Standardization of Biofilm Analysis Techniques

Advancements in the biofilm analysis methods allowed for better characterization, which, in turn, made it possible to achieve significant progress towards standardization. The first methods for identification and quantification of microorganisms in oral microcosms were based on denaturing gradient gel electrophoresis (DGGE) or checkerboard DNA–DNA hybridization. These methodologies could screen for a limited number of microbial species [24, 74]. More recent methodologies based on the identification of DNA with a large array of probes (human oral microbial identification microarray, HOMIM) or high-throughput direct identification of microbial species (next-generation sequencing based on massive parallel sequencing, HOMINGS) were able to identify virtually any microorganism that constitutes a biofilm community [75]. These latter methodologies also allowed to quantify the biodiversity of a biofilm and to assess shifts towards the prevalence of pathogenic species [76].

Due to the different bioreactor systems used and the increasing amount of biofilm data that is being gathered, there is a great need for standardization both of the bioreactor systems and of the biofilm analysis techniques, making possible direct comparisons among experiments differing in space (different research teams) and time. The first step in this direction has been made with the creation of two online platforms. The first one, MIABiE17 (minimum information about a biofilm experiment), is aimed to start providing guidelines about the minimum information that is to be acquired during an experiment involving biofilms. The other platform, BiofOmics18, is a systematic and standardized database that collects data about biofilm experiments.

4.5 Conclusions

Although a wide range of bioreactors are currently available, it is clear from this discussion that a universal bioreactor system that can be adapted to all clinical situations does not exist. Each model has its own advantages and limitations that must be acknowledged when choosing the model that best fits a distinct experimental design. There are some devices designed to study low fluid shear stresses, whereas others are more suitable for experiments under higher fluid shear stress. Some are appropriate when biofilm activity has to be evaluated, while other systems are better applied to the study of the biofilm structure. Furthermore, the operational flexibility of these models provides researchers with a spectrum of different models, often with overlapping characteristics. This situation, together with a reduced overall level of standardization, makes the comparison of the obtained results very difficult. Future studies in this field should be aimed at the standardization of the devices and analysis techniques.

From a materials science point of view, the science of biofilm development and bioreactor systems is most often difficult to be understood, given the high level of standardization that exists for the testing of the mechanical, physical, and chemical characteristics of a material. Furthermore, some research groups tend to use bioreactors as a "tool" to obtain simple answers about the antimicrobial activity of newly designed materials. While a quest for simplification of procedures and standardization of methods is always desirable, this approach often leads to undervaluing, or neglecting, many aspects that are intrinsic to the complexity of the material-host-biofilm interactions, and may lead to misinterpreting experimental results.

References

1. Simões M, Pereira MO, Sillankorva S, Azeredo J, Vieira MJ. The effect of hydrodynamic conditions on the phenotype of Pseudomonas fluorescens biofilms. Biofouling. 2007;23(3–4):249–58.
2. Marsh PD, Zaura E. Dental biofilm: ecological interactions in health and disease. J Clin Periodontol. 2017;44(S18):S12–22.
3. Peterson SN, Meissner T, Su AI, Snesrud E, Ong AC, Schork NJ, Bretz WA. Functional expression of dental plaque microbiota. Front Cell Infect Microbiol. 2014;4:108.
4. Senneby A, Davies J, Svensäter G, Neilands J. Acid tolerance properties of dental biofilms in vivo. BMC Microbiol. 2017;17(1):165.
5. Köves B, Magyar A, Tenke P. Spectrum and antibiotic resistance of catheter-associated urinary tract infections. GMS Infect Dis. 2017;5:Doc06.
6. Tenke P, Köves B, Nagy K, Hultgren SJ, Mendling W, Wullt B, Grabe M, Wagenlehner FM, Cek M, Pickard R, Botto H. Update on biofilm infections in the urinary tract. World J Urol. 2012;30(1):51–7.
7. Gomes IB, Meireles A, Gonçalves AL, Goeres DM, Sjollema J, Simões LC, et al. Standardized reactors for the study of medical biofilms: a review of the principles and latest modifications. Crit Rev Biotechnol. 2018;38(5):657–70.
8. Vickery K, Hu H, Jacombs AS, Bradshaw DA, Deva AK. A review of bacterial biofilms and their role in device-associated infection. Healthc Infect. 2013;18(2):61–6.
9. Costerton JW, Stewart PS, Greenberg EP. Bacterial biofilms: a common cause of persistent infections. Science. 1999;284(5418):1318–22.
10. Koo H, Allan RN, Howlin RP, Stoodley P, Hall-Stoodley L. Targeting microbial biofilms: current and prospective therapeutic strategies. Nat Rev Microbiol. 2017;15(12):740–55.
11. McBain AJ. Chapter 4: In vitro biofilm models: an overview. Adv Appl Microbiol. 2009;69:99–132.
12. Azeredo J, Azevedo N, Briandet R, Cerca N, Coenye T, Costa AR, Desvaux M, Di Bonaventura G, Hébraud M, Jaglic Z, Kačániová M. Critical review on biofilm methods. Crit Rev Microbiol. 2016;43(3):313–51.
13. Cazzaniga G. Resin-based composites modulate oral biofilm formation, PhD thesis. University of Milan, 2017. https://doi.org/10.13130/g-cazzaniga_phd2017-02-2314.

14. Marsh PD. In sickness and in health - what does the oral microbiome mean to us? An ecological perspective. Adv Dent Res. 2018;29(1):60–5.
15. Kroes I, Lepp PW, Relman DA. Bacterial diversity within the human subgingival crevice. Proc Natl Acad Sci U S A. 1999;96:14547–52.
16. Kawamura Y, Kamiya Y. Metagenomic analysis permitting identification of the minority bacterial populations in the oral microbiota. J Oral Biosci. 2012;54(3):132–7.
17. Ionescu A, Brambilla E, Hahnel S. Does recharging dental restorative materials with fluoride influence biofilm formation? Dent Mater. 2019;35(10):1450–63.
18. Hahnel S, Wastl DS, Schneider-Feyrer S, Giessibl FJ, Brambilla E, Cazzaniga G, Ionescu A. Streptococcus mutans biofilm formation and release of fluoride from experimental resin-based composites depending on surface treatment and S-PRG filler particle fraction. J Adhes Dent. 2014;16(4):313–21.
19. Nassar HM, Gregory RL. Biofilm sensitivity of seven Streptococcus mutans strains to different fluoride levels. J Oral Microbiol. 2017;9(1):1328265.
20. Chow J, Lee SM, Shen Y, Khosravi A, Mazmanian SK. Host–bacterial symbiosis in health and disease. Adv Immunol. 2010;107:243–74.
21. Kim Y-S, Kang S-M, Lee E-S, Lee JH, Kim B-R, Kim B-I. Ecological changes in oral microcosm biofilm during maturation. J Biomed Opt. 2016;21(10):101409.
22. Rudney JD, Chen R, Lenton P, Li J, Li Y, Jones RS, Reilly C, Fok AS, Aparicio C. A reproducible oral microcosm biofilm model for testing dental materials. J Appl Microbiol. 2012;113(6):1540–53.
23. Ionescu AC, Cazzaniga G, Ottobelli M, Garcia-Godoy F, Brambilla E. Substituted nano-hydroxyapatite toothpastes reduce biofilm formation on enamel and resin-based composite surfaces. J Funct Biomater. 2020;11(2):36.
24. Ledder RG, Gilbert P, Pluen A, Sreenivasan PK, Vizio WD, McBain AJ. Individual microflora beget unique oral microcosms. J Appl Microbiol. 2006;100(5):1123–31.
25. Robinson CJ, Bohannan BJM, Young VB. From structure to function: the ecology of host-associated microbial communities. Microbiol Mol Biol Rev. 2010;74(3):453–76.
26. Kuramitsu HK, He X, Lux R, Anderson MH, Shi W. Interspecies interactions within oral microbial communities. Microbiol Mol Biol Rev. 2007;71(4):653–70.
27. Sim CPC, Dashper SG, Reynolds EC. Oral microbial biofilm models and their application to the testing of anticariogenic agents. J Dent. 2016;50:1–11.
28. ten Cate JM. Models and role models. Caries Res. 2015;49(S1):3–10.
29. Rickard AH, Gilbert P, High NJ, Kolenbrander PE, Handley PS. Bacterial coaggregation: an integral process in the development of multispecies biofilms. Trends Microbiol. 2003;11(2):94–100.
30. Hajishengallis G, Liang S, Payne MA, Hashim A, Jotwani R, Eskan MA, McIntosh ML, Alsam A, Kirkwood KL, Lambris JD, Darveau RPA. Low-abundance biofilm species orchestrates inflammatory periodontal disease through the commensal microbiota and the complement pathway. Cell Host Microbe. 2011;10(5):497–506.
31. Nibali L, Henderson B. The human microbiota and chronic disease: dysbiosis as a cause of human pathology. Hoboken, NJ: John Wiley & Sons; 2016. p. 560.
32. Marsh PD, Hunter JR, Bowden GH, Hamilton IR, McKee AS, Hardie JM, et al. The influence of growth rate and nutrient limitation on the microbial composition and biochemical properties of a mixed culture of oral bacteria grown in a chemostat. J Gen Microbiol. 1983;129(3):755–70.
33. Shu M, Wong L, Miller JH, Sissons CH. Development of multispecies consortia biofilms of oral bacteria as an enamel and root caries model system. Arch Oral Biol. 2000;45(1):27–40.
34. McKee AS, McDermid AS, Ellwood DC, Marsh PD. The establishment of reproducible, complex communities of oral bacteria in the chemostat using defined inocula. J Appl Bacteriol. 1985;59(3):263–75.
35. Burmølle M, Webb JS, Rao D, Hansen LH, Sørensen SJ, Kjelleberg S. Enhanced biofilm formation and increased resistance to antimicrobial agents and bacterial invasion are caused by synergistic interactions in multispecies biofilms. Appl Environ Microbiol. 2006;72(6):3916–23.
36. Lin NJ. Biofilm over teeth and restorations: what do we need to know? Dent Mater. 2017;33(6):667–80.
37. Zanin ICJ, Gonçalves RB, Junior AB, Hope CK, Pratten J. Susceptibility of Streptococcus mutans biofilms to photodynamic therapy: an in vitro study. J Antimicrob Chemother. 2005;56(2):324–30.
38. Fernández CE, Tenuta LMA, Cury JA. Validation of a cariogenic biofilm model to evaluate the effect of fluoride on enamel and root dentine demineralization. PLoS One. 2016;11(1):e0146478.
39. Zhang A, Chen R, Aregawi W, He Y, Wang S, Aparicio C, et al. Development and calibration of biochemical models for testing dental restorations. Acta Biomater. 2020;109:132–41.
40. Forssten SD, Björklund M, Ouwehand AC. Streptococcus mutans, caries and simulation models. Nutrients. 2010;2(3):290–8.
41. Philip N, Suneja B, Walsh L. Beyond Streptococcus mutans: clinical implications of the evolving dental caries aetiological paradigms and its associated microbiome. Br Dent J. 2018;224(4):219.
42. Banas JA, Drake DR. Are the mutans streptococci still considered relevant to understanding the microbial etiology of dental caries? BMC Oral Health. 2018;18(1):129.
43. Matsumoto-Nakano M. Role of Streptococcus mutans surface proteins for biofilm formation. Jpn Dent Sci Rev. 2018;54(1):22–9.

44. Childers NK, Momeni SS, Whiddon J, Cheon K, Cutter GR, Wiener HW, et al. Association between early childhood caries and colonization with Streptococcus mutans genotypes from mothers. Pediatr Dent. 2017;39(2):130–5.
45. Hajishengallis E, Parsaei Y, Klein MI, Koo H. Advances in the microbial etiology and pathogenesis of early childhood caries. Mol Oral Microbiol. 2017;32(1):24–34.
46. Yu OY, Zhao IS, Mei ML, Lo EC-M, Chu C-H. Dental biofilm and laboratory microbial culture models for cariology research. Dent J. 2017;5(2):21.
47. Balouiri M, Sadiki M, Ibnsouda SK. Methods for in vitro evaluating antimicrobial activity: a review. J Pharm Anal. 2016;6(2):71–9.
48. Milho C, Andrade M, Boas DV, Alves D, Sillankorva S. Antimicrobial assessment of phage therapy using a porcine model of biofilm infection. Int J Pharm. 2019;557:112–23.
49. Morgan SJ, Lippman SI, Bautista GE, Harrison JJ, Harding CL, Gallagher LA, et al. Bacterial fitness in chronic wounds appears to be mediated by the capacity for high-density growth, not virulence or biofilm functions. PLoS Pathog. 2019;15(3):e1007511.
50. Bahamondez-Canas TF, Heersema LA, Smyth HD. Current status of in vitro models and assays for susceptibility testing for wound biofilm infections. Biomedicine. 2019;7(2):34.
51. Pierce CG, Uppuluri P, Tummala S, Lopez-Ribot JL. A 96 well microtiter plate-based method for monitoring formation and antifungal susceptibility testing of Candida albicans biofilms. JoVE J Vis Exp. 2010;44:e2287.
52. Skogman ME, Vuorela PM, Fallarero A. A platform of anti-biofilm assays suited to the exploration of natural compound libraries. JoVE J Vis Exp. 2016;118:e54829.
53. Kampf G. Antiseptic stewardship for wound and mucous membrane antiseptics. In: Kampf G, editor. Antiseptic stewardship: biocide resistance and clinical implications. Cham: Springer International Publishing; 2018. p. 689–94.
54. Ceri H, Olson ME, Stremick C, Read RR, Morck D, Buret A. The Calgary Biofilm Device: new technology for rapid determination of antibiotic susceptibilities of bacterial biofilms. J Clin Microbiol. 1999;37(6):1771–6.
55. Brown JL, Johnston W, Delaney C, Short B, Butcher MC, Young T, et al. Polymicrobial oral biofilm models: simplifying the complex. J Med Microbiol. 2019;68(11):1573–84.
56. Bradshaw DJ, Marsh PD, Schilling KM, Cummins D. A modified chemostat system to study the ecology of oral biofilms. J Appl Bacteriol. 1996;80(2):124–30.
57. Peters AC, Wimpenny JWT. A constant-depth laboratory model film fermentor. Biotechnol Bioeng. 1988;32(3):263–70.
58. Rozenbaum RT. Antimicrobial and nanoparticle penetration and killing in infectious biofilms, PhD thesis. Rijksuniversiteit Groningen, 2019. http://hdl.handle.net/11370/0f2d1f8e-8898-4fb3-af42-7e8fd68c58e5.
59. Cotter JJ, O'Gara JP, Stewart PS, et al. Characterization of a modified rotating disk reactor for the cultivation of Staphylococcus epidermidis biofilm. J Appl Microbiol. 2010;109:2105–17.
60. Möhle RB, Langemann T, Haesner M, Augustin W, Scholl S, Neu TR, et al. Structure and shear strength of microbial biofilms as determined with confocal laser scanning microscopy and fluid dynamic gauging using a novel rotating disc biofilm reactor. Biotechnol Bioeng. 2007;98(4):747–55.
61. Kharazmi A, Giwercman B, Høiby N. Robbins device in biofilm research. In: Methods in enzymology. New York: Academic Press; 1999. p. 207–15.
62. Jass J, Costerton JW, Lappin-Scott HM. Assessment of a chemostat-coupled modified Robbins device to study biofilms. J Ind Microbiol. 1995;15(4):283–9.
63. Goeres DM, Hamilton MA, Beck NA, Buckingham-Meyer K, Hilyard JD, Loetterle LR, et al. A method for growing a biofilm under low shear at the air–liquid interface using the drip flow biofilm reactor. Nat Protoc. 2009;4(5):783–8.
64. Ledder RG, McBain AJ. An in vitro comparison of dentifrice formulations in three distinct oral microbiotas. Arch Oral Biol. 2012;57(2):139–47.
65. Ionescu A, Wutscher E, Brambilla E, Schneider-Feyrer S, Giessibl FJ, Hahnel S. Influence of surface properties of resin-based composites on in vitro Streptococcus mutans biofilm development. Eur J Oral Sci. 2012;120(5):458–65.
66. Brambilla E, Ionescu A, Cazzaniga G, Edefonti V. The influence of antibacterial toothpastes on in vitro Streptococcus mutans biofilm formation: a continuous culture study. Am J Dent. 2014;27(3):7.
67. Yoon HY, Lee SY. Establishing a laboratory model of dental unit waterlines bacterial biofilms using a CDC biofilm reactor. Biofouling. 2017;33(10):917–26.
68. Li Y, Carrera C, Chen R, Li J, Lenton P, Rudney JD, et al. Degradation in the dentin–composite interface subjected to multispecies biofilm challenges. Acta Biomater. 2014;10(1):375–83.
69. Yawata Y, Nguyen J, Stocker R, Rusconi R. Microfluidic studies of biofilm formation in dynamic environments. In: O'Toole GA, editor. J Bacteriol. 2016;198(19):2589–95.
70. Benoit MR, Conant CG, Ionescu-Zanetti C, Schwartz M, Matin A. New device for high-throughput viability screening of flow biofilms. Appl Environ Microbiol. 2010;76(13):4136–42.
71. Groisman A, Lobo C, Cho H, Campbell JK, Dufour YS, Stevens AM, et al. A microfluidic chemostat for experiments with bacterial and yeast cells. Nat Methods. 2005;2(9):685–9.
72. Kim J, Hegde M, Kim SH, Wood TK, Jayaraman A. A microfluidic device for high throughput bacterial biofilm studies. Lab Chip. 2012;12(6):1157–63.
73. Busscher HJ, van der Mei HC. Microbial adhesion in flow displacement systems. Clin Microbiol Rev. 2006;19(1):127–41.

74. Gellen LS, Wall-Manning GM, Sissons CH. Checkerboard DNA-DNA hybridization technology using digoxigenin detection. In: Hilario E, Mackay J, editors. Protocols for nucleic acid analysis by nonradioactive probes. Totowa, NJ: Humana Press; 2007. p. 39–67.
75. Mougeot J-LC, Stevens CB, Cotton SL, Morton DS, Krishnan K, Brennan MT, et al. Concordance of HOMIM and HOMINGS technologies in the microbiome analysis of clinical samples. J Oral Microbiol. 2016;8(1):30379.
76. Adams SE, Arnold D, Murphy B, Carroll P, Green AK, Smith AM, et al. A randomised clinical study to determine the effect of a toothpaste containing enzymes and proteins on plaque oral microbiome ecology. Sci Rep. 2017;7(1):43344.

Surface Properties of Dental Materials and Biofilm Formation

5

Ralf Bürgers, Sebastian Krohn, and Torsten Wassmann

Abstract

Bacterial adhesion to biological tissues of the oral cavity or artificial dental materials and the subsequent formation of complex biofilms are responsible for major dental pathologies such as caries, periodontitis, peri-implantitis, denture stomatitis, and candidiasis [1]. The first and essential step in biofilm formation is the initial attachment of single microbes to a substratum, where they have to interact with the available physicochemical surface conditions in order to remain and multiply [2, 3]. As a matter of principle, bacteria exist naturally within structured communities growing as biofilms and sufficient bacterial adhesion in the oral cavity is therefore the only way to survive for most bacteria in the long run [2, 4].

5.1 Introduction

Bacterial adhesion to biological tissues of the oral cavity or artificial dental materials and the subsequent formation of complex biofilms are responsible for major dental pathologies such as caries, periodontitis, peri-implantitis, denture stomatitis, and candidiasis [1]. The first and essential step in biofilm formation is the initial attachment of single microbes to a substratum, where they have to interact with the available physicochemical surface conditions in order to remain and multiply [2, 3]. As a matter of principle, bacteria exist naturally within structured communities growing as biofilms and sufficient bacterial adhesion in the oral cavity is therefore the only way to survive for most bacteria in the long run [2, 4].

Bacterial attachment on dental materials is dependent on several interlocking factors between the abiotic substratum and the biotic cell surface of the bacteria (see Fig. 5.1).

In detail, these are properties of the bacterium (such as the varying potential for specific interactions between ligands and receptors or different surface-bound proteins of the cell surface), environmental conditions (for example composition of the saliva, pH, and metabolites), and physicochemical properties of the target substratum [2, 5–7]. Surface properties that have been discussed to influence the quantity and quality of bacterial adhesion are [8]:

1. Surface roughness, topography
2. Surface energy, hydrophobicity
3. Surface charge, zeta potential
4. Substratum chemistry
5. Substratum stiffness

R. Bürgers (✉) · S. Krohn · T. Wassmann
Department of Prosthodontics, University Medical Center Göttingen, Georg August University Göttingen, Göttingen, Germany
e-mail: ralf.buergers@med.uni-goettingen.de; sebastian.krohn@med.uni-goettingen.de; torsten.wassmann@med.uni-goettingen.de

A. C. Ionescu, S. Hahnel (eds.), *Oral Biofilms and Modern Dental Materials*,
https://doi.org/10.1007/978-3-030-67388-8_5

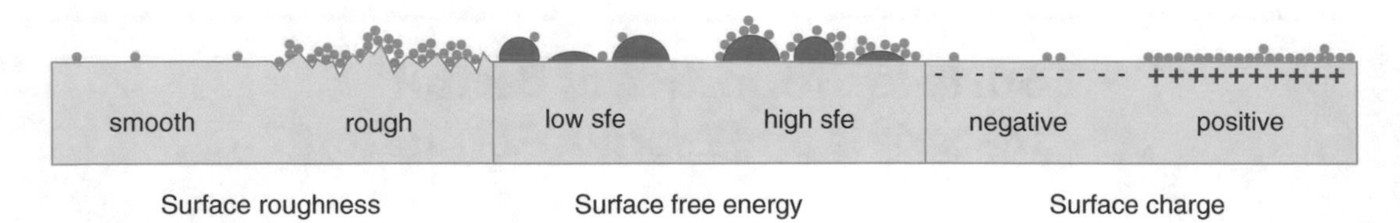

Fig. 5.1 Schematic overview of bacterial adhesion to solid surface substrata: the potential to adhere bacterial cells is represented by the amounts of adhering bacteria (orange)

For both dentists working in practice and scientists focused on oral biofilm topics, it is crucial to understand how these surface properties (of dental materials) affect microbe-surface interactions in order to control biofilm-related oral infections [8].

5.2 Surface Roughness and Topography

The topography of a real solid surface is defined by spatial frequencies as deviations of its profile from an ideal (i.e., totally flat) surface. These frequencies are classified into three groups depending on the value of the irregularity step: the high frequency (small amplitude) is referred to as roughness, the medium frequency (large amplitude) as waviness, and the low frequency (macroscopic) as form (see Fig. 5.2) [7, 9]. These three parameters may be separated by filtering; however roughness is the most important regime in context with microbial adhesion and biofilm formation. When investigating the influence of the substratum profile on adhesion of bacterial cells, characterization of the substratum topography is focused on micron and submicron levels of roughness [7].

Simply said, a rough and therefore structured surface is a folded smooth surface with exactly the same interaction potential to adhering bacteria, but with an extended surface area [6, 10]. Thus, roughening of materials increases the surface area (and therefore the number of possible interaction contacts) by a factor of 2–3 [4]. The higher the number of available attachment points on a given surface area, the more the cell attachment [3, 11]. Additionally to the absolute number of adhering bacteria, the average adhesion force per bacterial cell may rise with increasing roughness of the surface, which results from enhancing the quantity of available contact areas per cell on the substratum (see Fig. 5.3). This theoretical consideration might explain why "optimal" surface roughness values are often defined in relation to specific cell sizes of bacteria [3, 11].

Scratches and grooves are no preferred areas for initial bacterial adhesion per se, but in the oral cavity, microbial cells are very well protected from ubiquitous shear forces in these sites [6, 12]. Therefore, bacterial adhesion starts from surface irregularities, where they are sheltered against forces of removal (salivary flow, sulcus fluid, nutrition, oral hygiene, and natural movement of the soft tissues). Scanning electron microscopy exhibited that initial bacterial adhesion of hard tooth tissues mostly starts from small surface defects such as cracks, grooves, or fissures [4, 13, 14].

5.2.1 Characterization Techniques for Surface Roughness and Topography

In general, surface profile characterization may be performed with different types of profilometers, which can be divided into contact-type or tactile measuring instruments (mechanical stylus profilometer/perthometer, atomic force microscope (AFM), scanning tunneling microscopy) and noncontact-type or optical measuring instruments (white light interferometer, laser scanning confocal microscope, widefield confocal microscope) [3, 7, 15–19].

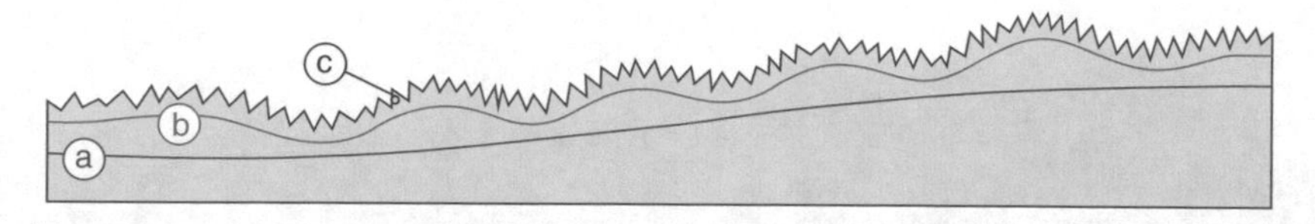

Fig. 5.2 Levels of topography on a 2D measured profile of a solid surface, separated as nominal form (a, red), waviness (b, green), and roughness (c, blue)

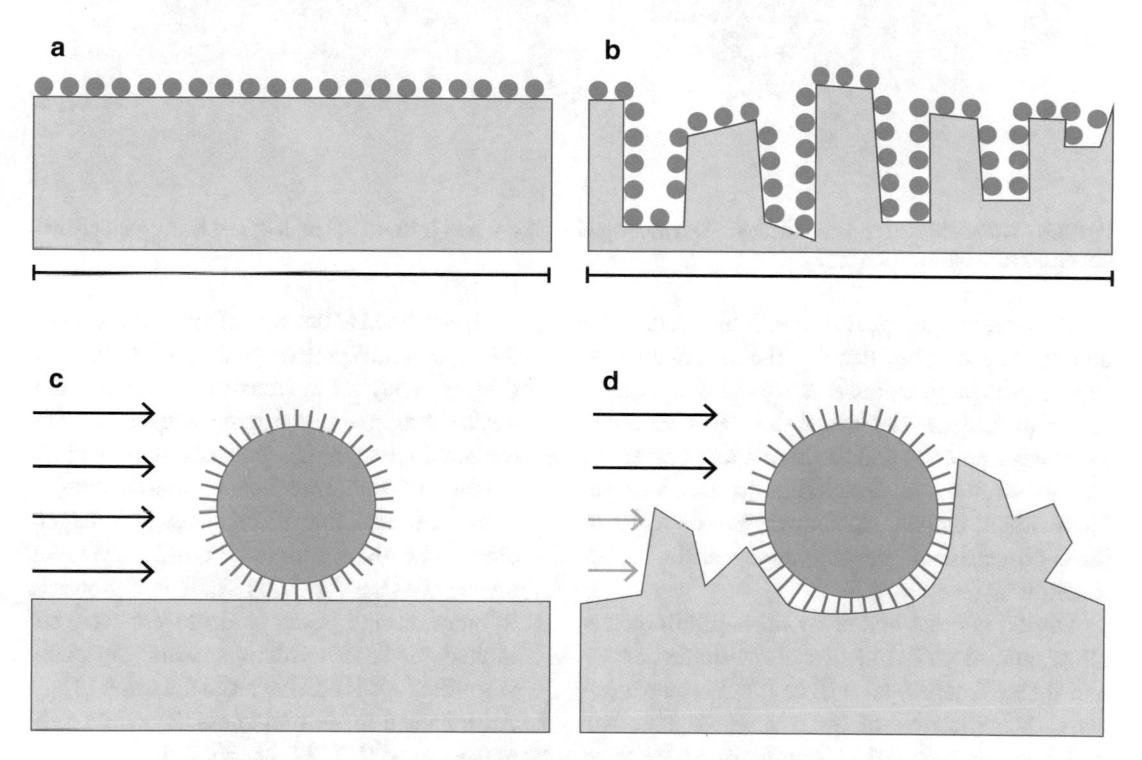

Fig. 5.3 Lower number of bacteria on a flat surface (**a**; $n = 20$); roughening increases the number of adhering cells (**b**; $n = 50$). Roughening of the surface results in larger contact areas between bacteria and substratum and less effective shear forces (arrows) (**c**, **d**)

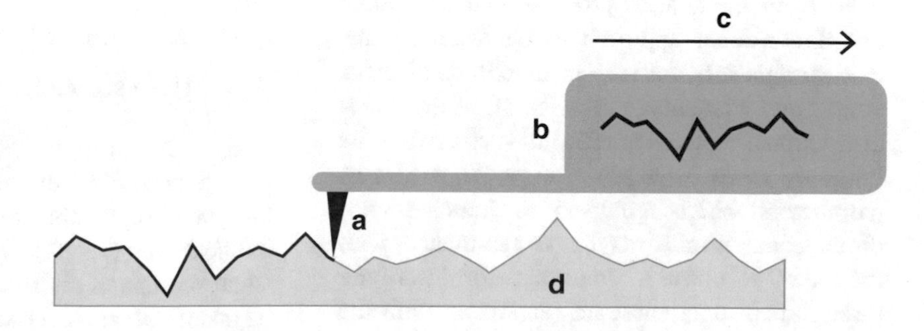

Fig. 5.4 Contact-type surface roughness measuring a stylus (a) is connected to a detector (b) and moved laterally (c) in contact with a solid surface sample (d)

Contact profilometers consist of a detector tip with a stylus (diamond or sapphire, diameter about 10 nm), which is moved in contact laterally along the surface sample for a specific distance and a specified contact force. The vertical motion of the stylus is electrically detected, and the changes in the height (Z position) of the arm holder are displayed to reconstruct the surface profile (see Fig. 5.4).

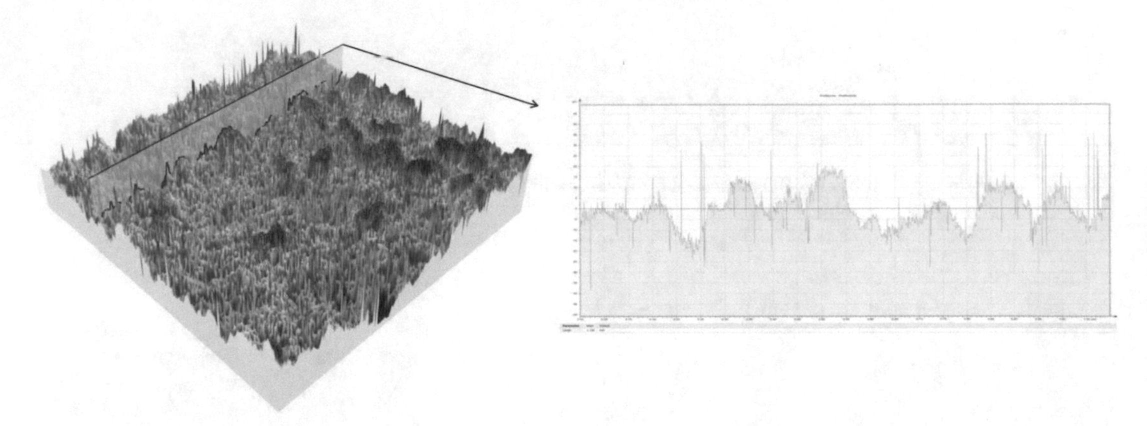

Fig. 5.5 Widefield confocal microscopy: 3D rendering of a surface area (110 × 110 μm, left) and corresponding two-dimensional profile section (right)

This technique provides reliable data with adequate resolution, because the stylus directly touches the sample, but it is slower than optical techniques, shape and size of the stylus influence the measurements, and the permanent contact to the substratum is destructive to the surface. Noncontact optical techniques are capable of three-dimensional measurements with higher resolutions (see Fig. 5.5).

Digital contact and noncontact profilometers allow automated calculation of multitude of two- and three-dimensional surface roughness parameters. Modification of test materials (e.g., via polishing or coating) or comparisons between different samples may be characterized by visualization (2D and 3D) and calculation of surface roughness parameters (see Fig. 5.6).

In general, roughness parameters are calculated from the filtered profiles. The traditional two-dimensional approach to characterize surface roughness by the average amplitude of peaks is not fully adequate to describe all of the three-dimensional features of all solid surfaces and its influence on bacterial adhesion [8, 20, 21]. Most parameters which are used to indicate two-dimensional profile roughness (annotated with the letter "*R*") have a counterpart for describing three-dimensional substrata (annotated with the letter "*S*") [7, 22]. Exemplary and frequently used roughness/surface parameters are given in the subsequent paragraph:

1. Maximum roughness = maximum height of the profile (R_t or R_{max}) [22–24]

   R_t is defined as the vertical distance between the maximum profile peak height and the deepest valley (or the maximum profile valley depth) within the sampling length (*l*). The evaluation length (*L*) consists of a defined number of sampling length, mostly five or more. The mean line of roughness *M* is the reference line about which the profile deviations are located (see Fig. 5.7) [23]. R_t is appropriate to indicate high peaks or deep scratches, but limited for further interpretations, especially when the focus is on bacterial adhesion [22].

2. Arithmetical mean roughness value = roughness average (R_a) [22, 23, 25, 26]

   R_a is the average absolute deviation of all roughness irregularities (peaks and valleys y_i) from *M* within *l* [22]; it indicates the arithmetical mean deviation from *M* (see Fig. 5.8) [26].

   R_a is calculated from the equation $R_a = \frac{1}{l}\int_0^l |y(x)|\,dx$ [22, 24].

   R_a is the most frequently used roughness parameter in literature and therefore it should be given for comparisons between different publications [7, 22]. Nevertheless, R_a has some significant disadvantages which have to be taken into consideration when applied and interpreted. It does not give any information about the spatial distribution of peaks and valleys (i.e., of the wavelength) and it may not indicate small differences in various test surfaces [7, 22]. Additionally R_a does not differ-

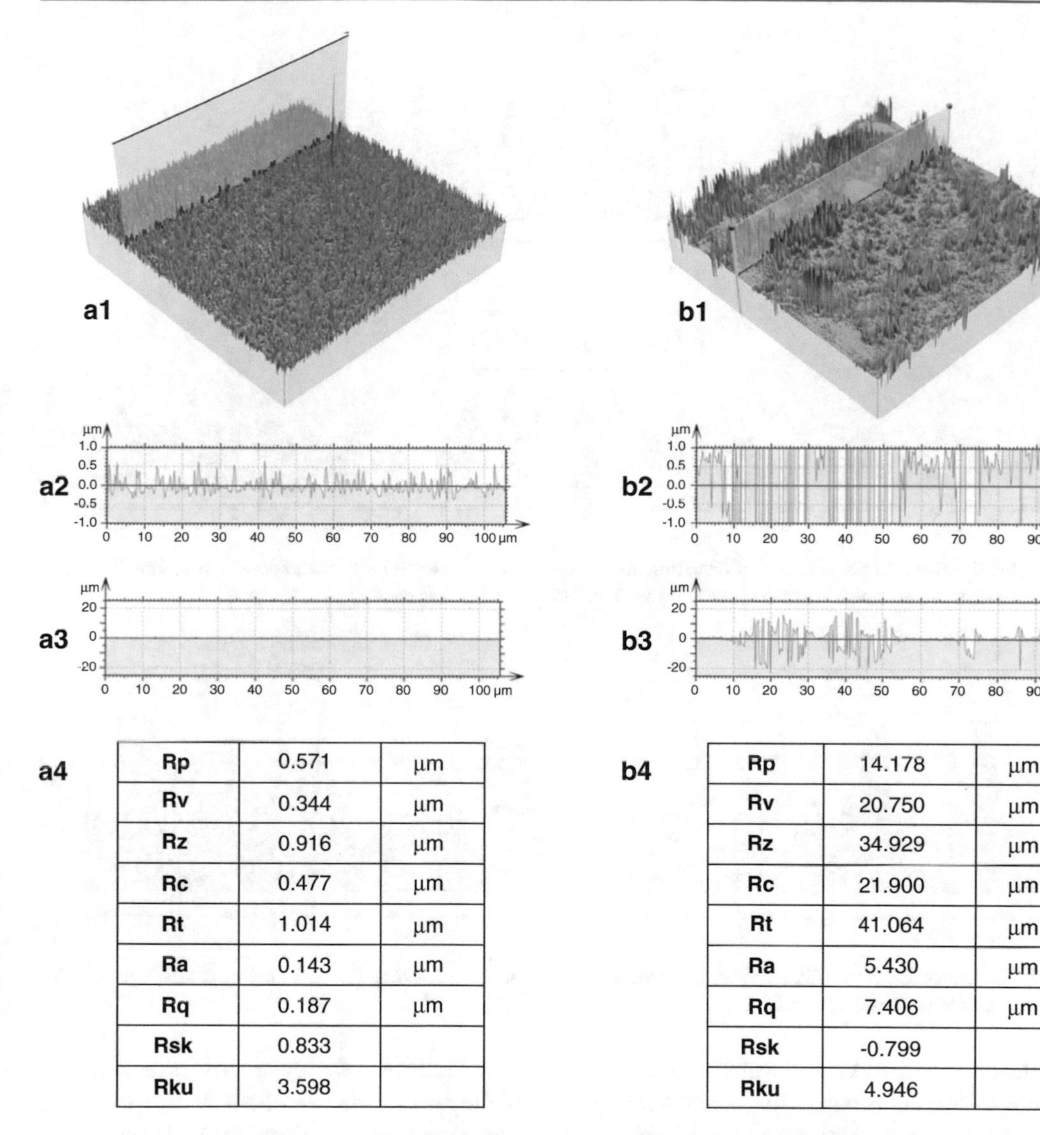

a4

Rp	0.571	µm
Rv	0.344	µm
Rz	0.916	µm
Rc	0.477	µm
Rt	1.014	µm
Ra	0.143	µm
Rq	0.187	µm
Rsk	0.833	
Rku	3.598	

b4

Rp	14.178	µm
Rv	20.750	µm
Rz	34.929	µm
Rc	21.900	µm
Rt	41.064	µm
Ra	5.430	µm
Rq	7.406	µm
Rsk	-0.799	
Rku	4.946	

Fig. 5.6 Surface roughness analysis via widefield confocal microscopy: comparison of composite direct filling materials after pre-polishing (**a**) and after sandblasting (**b**; Al_2O_3, 110 µm, 2 bar). Reports of 3D profiles 110 × 110 µm (**a**$_1$ and **b**$_1$), 2D profile *Y*-axis ± 1 µm (**a**$_2$ and **b**$_2$), 2D profile *Y*-axis ± 25 µm (**a**$_3$ and **b**$_3$), and automatically calculated roughness parameters (**a**$_4$ and **b**$_4$)

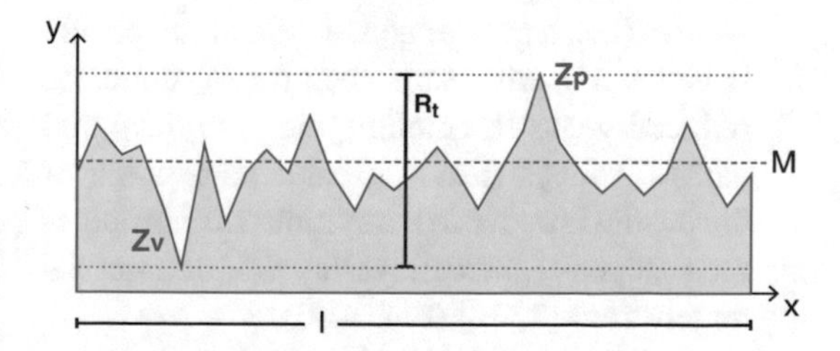

Fig. 5.7 Maximum roughness R_t/R_{max}, mean line of roughness (*M*), and sampling length (*l*)

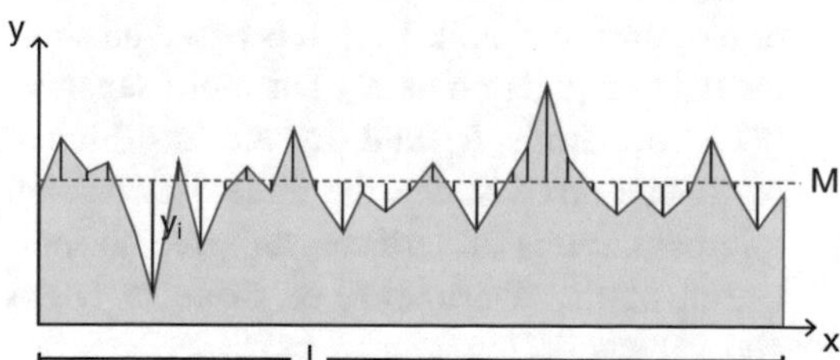

Fig. 5.8 Arithmetical mean roughness R_a and deviations y_i from mean line (*M*) over sampling length (*l*)

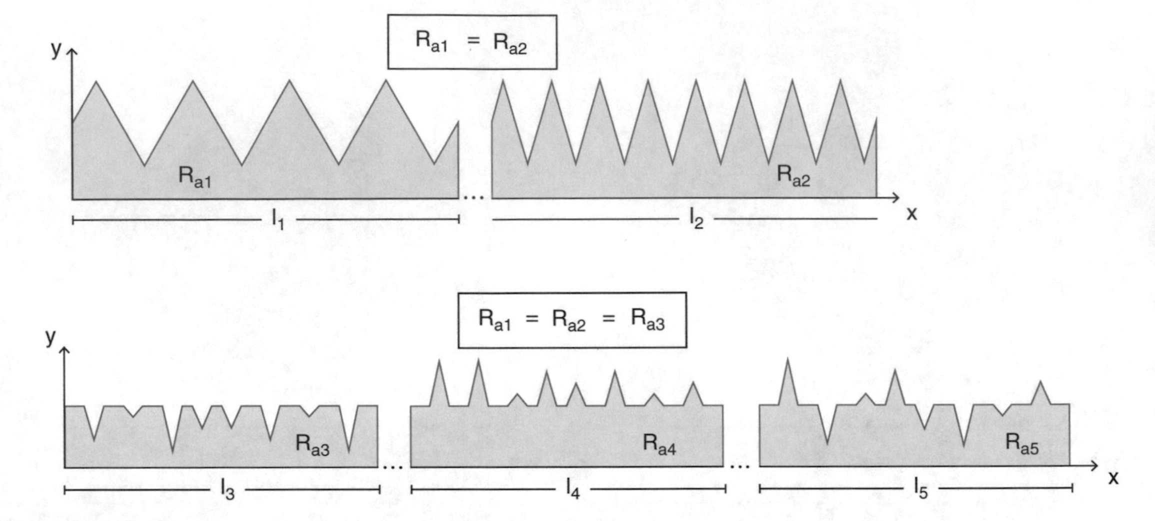

Fig. 5.9 The R_a values do not give any information about the intervals between heights and depth (therefore $R_{a1} = R_{a2}$), and the R_a does not distinguish between peaks and valleys (therefore $R_{a1} = R_{a2} = R_{a3}$)

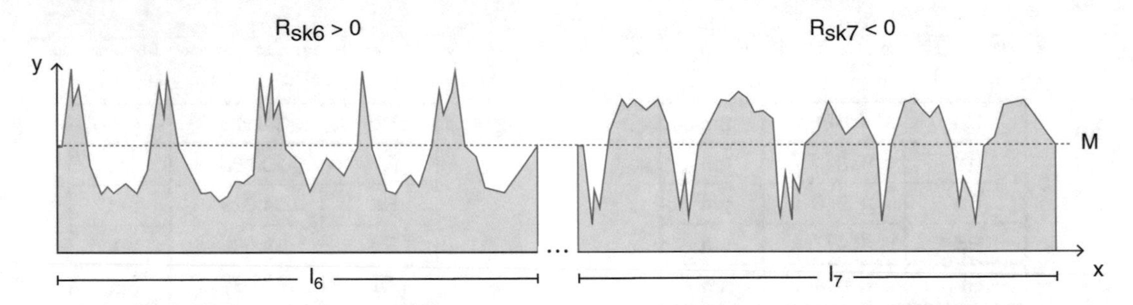

Fig. 5.10 Skewness R_{sk} of two different surface profiles with same R_a ($R_{a6} = R_{a7}$). For l_6, a greater percentage is below M; for l_7 the profile is mainly above M

entiate between peaks and valleys and may therefore not be suitable for characterizing variances between different types of surface profiles (see Fig. 5.9) [26].

3. Root mean square roughness (R_q or RMS) [22, 23, 26, 27]

   R_q is the standard deviation of the distribution of profile heights [22]. It is based on similar data acquisition as R_a, but more sensitive [22, 26]. Both R_a and R_z are insufficient parameters to describe the three-dimensional microstructures of surfaces and give no indication of the distribution or shape of peaks and valleys [27].

   R_q is calculated from the equation $R_q = \sqrt{\frac{1}{l}\int_0^l \{y(x)\}^2\,dx}$ [22, 26].

4. Skewness (R_{sk}) [22, 23]

   R_{sk} indicates the symmetry or asymmetry of the surface profile about M and therefore overcomes one significant limitation of R_a and R_q. In practice, it may specify differences in the shape of two or more surface profiles with the same R_a and R_q [22]. A positive R_{sk} indicates higher peaks with shallower, broader valleys (i.e., a greater percentage of the profile is below M), whereas a negative R_{sk} indicates reduced peaks in combination with deep and narrow valleys (i.e., a greater percentage of the profile is above M); a symmetrical arrangement of peaks and valleys has a R_{sk} of approximately zero [7, 22] (Fig. 5.10).

   R_{sk} is calculated from the equation $R_{sk} = \frac{1}{R_q^3}\int_{-\infty}^{\infty} y^3 p(y)\,dy$ [22].

5.2.2 Surface Roughness, Topography, and Biofilm Formation

Innumerable studies attempted to extract the nature of the interactions between surface roughness and microbial adhesion. The detailed biological and physicochemical mechanisms by which topographic parameters modulate bacterial and fungal attachment remain (at least partially) unclear [7, 15, 28–30].

Most authors nevertheless suggest that attachment of bacteria/biofilm is directly correlated to surface roughness and the clinical observation through the eyes of a practical dentist shows that smooth and highly polished surfaces reduce plaque formation on most dental materials in the oral cavity [31–34]. Higher quantities of adhering microorganisms on rougher surfaces have been shown in vitro [35–37] and in vivo [38–41]. Teughels and coworkers concluded in a review based on 24 in vivo studies that dental biofilm accumulates and retains more on rougher surfaces in terms of biofilm thickness, colonized area, and colony-forming units (CFU) [42].

Additionally, it has been shown that smoothening of a solid surface (for example by polishing) can reduce biofilm formation [43]. On titanium (implant) surfaces, smoothening below 0.2 µm (R_a) showed no further reducing effect on the adhering biofilm, either in the quantity of bacteria or in the pathogenicity of adhering bacteria. This specific value has later often been cited as a threshold surface roughness below which "bacterial adhesion cannot be reduced further" [44–46].

The correlation between surface irregularities (i.e., surface roughness) and initial microbial attachment leading to biofilm formation may be explained by three main factors [3, 14, 42]:

1. The initial adhesion of single early-colonizing bacteria to solid surfaces starts at areas where these single cells are better protected against oral shear forces, which in turn gives them the adequate time to change from reversible attachment to irreversible and specific adhesion (see Fig. 5.3d) [14, 42].
2. Roughening of materials per se increases the available surface area, which then increases the absolute number of adhering cells and the adhesion force per bacterium (see Fig. 5.6) [3].
3. Removal of bacteria and biofilms is much more difficult on micro- and macrostructured surfaces, resulting in faster and more intensive recolonization [14, 42].

In any investigation concerning surface roughness and biofilm formation, correlations may not only be reduced to pure roughness or topographic parameters. Quantity and quality of bacterial adhesion on solid surfaces are dependent not only on the physicochemical properties of the substratum, but also on the cell size, shape, and composition of the surface of the bacterial species in the surrounding [8]. All hard and soft tissues of the oral cavity and all dental materials exposed to the oral biotope are immediately covered by the salivary pellicle. This protein layer is up to 1000 nm thick with great interindividual differences and additionally alters the nanotopography, microstructure, and surface roughness of the solid substratum [8, 47].

5.2.3 Interaction Between Surface Roughness (or Topography) and Surface Free Energy (or Hydrophobicity)

Both surface free energy and surface roughness have a major impact on the attachment of oral microorganisms on dental substrata. The influence of surface roughness (or surface topography) seems to be predominant towards surface free energy and hydrophobicity in the context of microbial adhesion and biofilm formation (on dental materials) [4, 42, 48, 49].

Additionally, both surface properties influence each other, whereby surface topography is a major determinant of surface free energy and wettability (i.e., hydrophobicity and hydrophilicity) [27, 50]. Surface roughening will decrease contact angles if the initial contact angle is below 60°, whereas surface roughening will further increase contact angles on substrata with initial contact angles above 86°. On surfaces with initial contact angles between 60° and 86° surface roughening or smoothing will not affect contact angles, just like changes in R_a below 0.1 µm do not affect contact angles at all and for this reason additionally not surface free energies [4, 42, 48, 51].

5.3 Wettability (i.e., Hydrophobicity or Hydrophilicity) and Surface Free Energy (SFE)

Wettability (=wetting) is the potential of a liquid to maintain contact with a solid substratum, resulting from different intermolecular interactions between both surfaces, and it reflects how a specific liquid behaves on a solid surface. If the liquid is water, surfaces which repel water are labeled as hydrophobic, whereas hydrophilic surfaces are attracted to water.

Wettability is quantified by the contact angle (θ, theta, [°]), which is the angle between the surface of the liquid and the surface of the solid. Complete wetting (i.e., the droplet spreads completely) results in a contact angle of 0°, between 0° and 90° the surface is wettable (hydrophilic), above 90° it is not wettable (hydrophobic), and the theoretical limit of θ is 180° with the droplet standing on the surface (see Fig. 5.11) [3]. Superhydrophilic surfaces are those with $C < 5°$, and superhydrophobic are those with $\theta > 150°$ [3, 52].

If the test substratum is smooth, rigid, and insoluble, the contact angle θ can be defined by Young's equation (see Fig. 5.12) [53, 54]:

$$\cos\theta = \frac{\gamma_{sv} - \gamma_{sl}}{\gamma_{lv}}$$

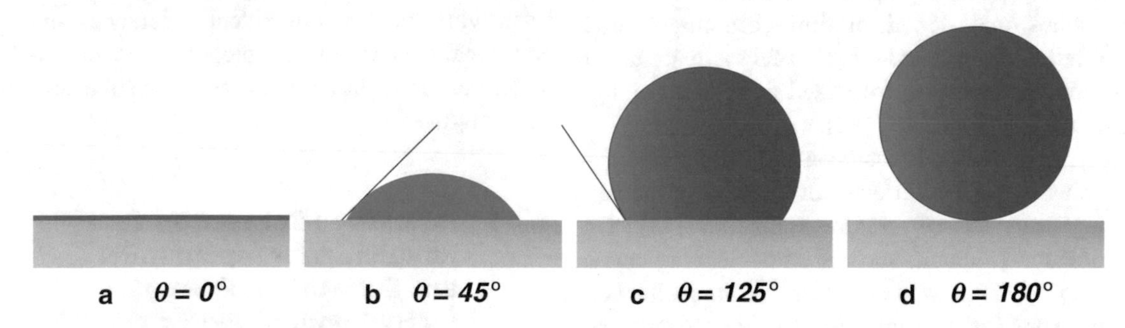

Fig. 5.11 Contact angles on different substrata: complete wetting on superhydrophilic surface (**a**), good wetting on hydrophilic surface (**b**), bad wetting on hydrophobic surface (**c**), and non-wetting on superhydrophobic surface (**d**)

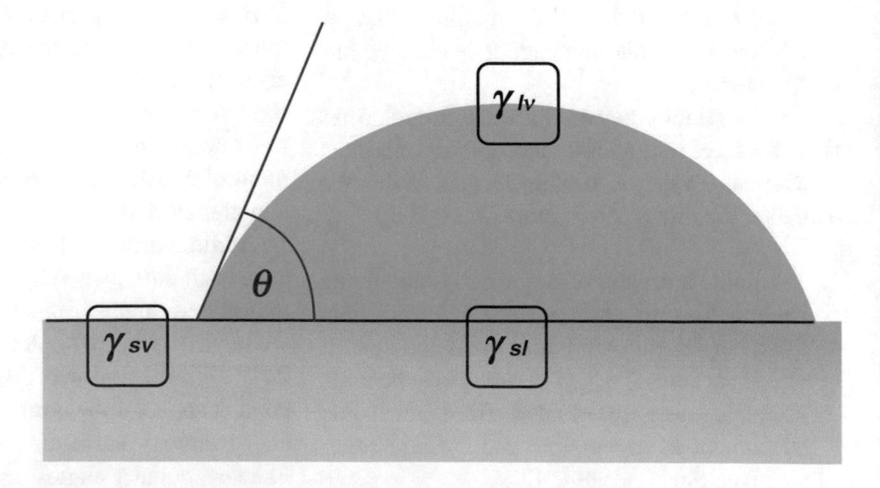

Fig. 5.12 Contact angles are related to the surface free energies according to the equation of Young

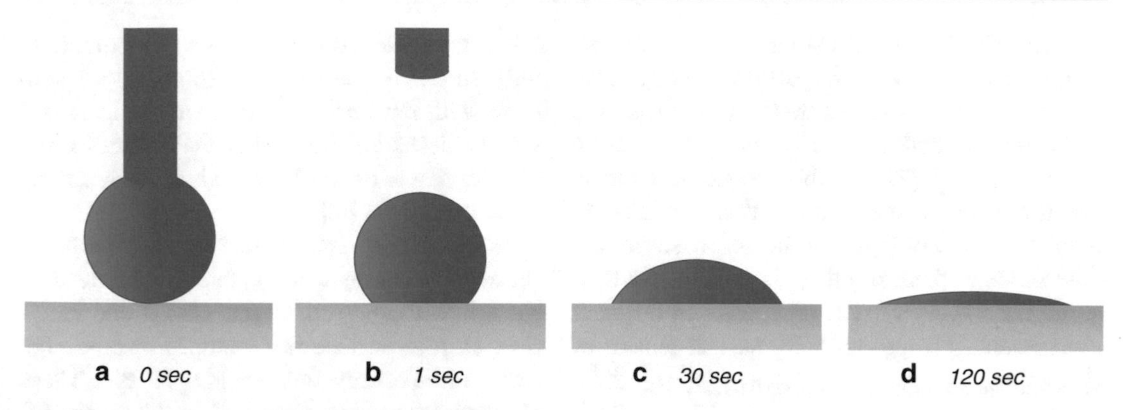

Fig. 5.13 Water contact angle measurement: a droplet is released through a microliter syringe (**a**), syringe is removed (**b**), time-dependent change of contact angle (**c**, **d**)

- θ: Young's contact angle
- γ_{sv}: SFE between solid (s) and vapor (v)
- γ_{sl}: SFE between solid (s) and liquid (l)
- γ_{lv}: SFE between liquid (l) and vapor (v)

$\gamma \left(\text{gamma}, \left[\frac{\text{N}}{\text{m}} = \frac{\text{J}}{\text{m}^2} \right] \right)$ is the surface tension (or surface energy or surface free energy) and is measured in force per unit length (or energy per unit area). It is the amount of energy, necessary to break chemical bonds [3]. While surface tension is used relating to liquids, surface energy is used relating to solids [3].

The surface free energy of solid substrata may be separated in a dispersive (*d*) and a polar (*p*) component (according to the Owens and Wendt geometric mean approach) [3, 52, 55]:

$$\gamma_{\text{L}}(1+\cos\theta) = 2\left[\left(\gamma_{\text{L}}{}^{p}\gamma_{\text{S}}{}^{p}\right)^{1/2} + \left(\gamma_{\text{L}}{}^{d}\gamma_{\text{S}}{}^{d}\right)^{1/2}\right]$$

γ_{L} is the liquid surface tension and γ_{S} is the solid surface free energy.

5.3.1 Characterization Techniques for Hard Surface Wettability and SFE

In practice, contact angles θ are measured with the relatively simple sessile droplet method (contact angle measurement, CAM). In general, a droplet of a liquid is placed on a planar solid surface with a microliter syringe and the angle between the surface of the liquid and the solid is analyzed (see Fig. 5.11) [3]. Mostly, water CAM by using a goniometer is used to determine hydrophobicity/hydrophilicity. The results of goniometer measurements may be used to calculate the surface energy of the test solid via Young's equation [6, 15, 56, 57]. CAM is time dependent; normally the water droplet will flatten on the solid surface; that is, the contact angle will decrease over time (see Fig. 5.13). By using three liquids with different wettability and known surface tension components (γ_{L}, $\gamma_{\text{L}}{}^{p}$, $\gamma_{\text{L}}{}^{d}$), the surface free energy of any test solid with its polar and dispersion component can be measured by contact angle measurements and then calculated according to the Owens, Wendt, Rabel, and Kaelble method and based on Young's equation [52, 55, 58–63].

5.3.2 Characterization Techniques for Cell Surface Wettability (and SFE)

Cell surface hydrophobicity cannot be measured directly, but the interaction between (bacterial) cells and hydrophobic/hydrophilic hard surfaces may be used as an index for cell wettability [15]. For example, bacterial attachment to hydrocarbon (BATH) [64–66], hydrophobic interaction chromatography (HIC) [67], salting-out aggregation test (SAT) [68], CAM on a lawn of (bacterial) cells [58, 69, 70], and partitioning of cells in two-phase systems (TPP) are used for measuring cell surface hydrophobicity (summarized by Rosenberg et al. [71]). In this context, it has to be mentioned that bacteria are vital cells and adapt their surface composition in response to the environment; therefore there is no constant SFE value for a specific bacterial species in different experimental settings [14]. The surface tension of streptococcal bacteria is approximately $100\ \frac{\mathrm{erg}}{\mathrm{cm}^2} = 10\ \frac{0{,}1\,\mu J}{\mathrm{cm}^2}$ [69, 70].

5.3.3 Wettability, Surface Free Energy, and Biofilm Formation

The influence of hydrophobicity and surface free energy on microbial attachment and biofilm formation has been demonstrated in vitro and in vivo [3, 72–75]. Additionally, it has been shown in various experimental settings that bacterial adhesion can be either promoted or inhibited by tuning hydrophobicity and surface free energy [8, 76]. Therefore, many scientific approaches have been proposed to create anti-adhesive and antimicrobial materials by modifying wettability or SFE properties [77, 78].

Various scientific strategies have been proposed to minimize biofilm formation by alterations of surface free energy or hydrophobicity [77, 78].

A review by Teughels et al. showed that generally solid surfaces with a higher surface free energy are preferred by adhering bacteria [42]. These observations have been explained by weaker interactions between bacteria and surface, caused by insufficient binding to the conditioning layer [42, 79, 80]. In the oral cavity, all surfaces will instantly become conditioned by the acquired pellicle with low surface tension, which mostly results in a reduction of the quantity of adhering bacteria. The influence of solid hydrophobicity and surface free energy on initial bacterial colonization is transferred through the salivary pellicle, even in the in vivo/in situ setting [42]. In detail, the surface free energy of substrata with low surface energy will be increased with pellicle formation, while the surface energy of substrata with initially high surface energy decreases [4, 81, 82].

Some authors reported that substrata with low surface free energy were preferably colonized by bacteria with low surface free energy and bacteria with high surface free energies favored high surface free energy surfaces [42, 77, 83]. These observations may be explained by the thermodynamic approach and the equation of Dupré ($\Delta F^{\mathrm{adh}} = \gamma_{\mathrm{BS}} - \gamma_{\mathrm{BL}} - \gamma_{\mathrm{SL}}$) [4]. Most bacteria have high surface energies and saliva has a relatively low surface tension. Therefore, bacterial adhesion is enhanced by high solid surface free energies [4]. Additionally, substrata with low surface free energy may then be colonized in a higher percentage by microorganisms with lower surface free energies [4]. The preference of hydrophobic or hydrophilic substrata differs among the various microbial species [8, 77, 78]. Nevertheless, distinct and universally valid correlations between bacterial adhesion and solid surface free energies or material-specific antimicrobial strategies cannot be concluded, because various other influencing factors such as topography and chemical composition of the substratum, composition of the conditioning layer (saliva, sulcus fluid, blood, …), and changes in the surrounding vapor cannot be excluded and may be predominant [4].

Numerous studies showed that superhydrophobic and superhydrophilic surfaces may both be used to prevent oral biofilm formation [3]. Superhydrophobicity is inspired by the self-cleaning (better easy-to-clean) property of the lotus leaf with water contact angle of approximately 170° (see Fig. 5.14) [3, 84]. It is used to limit the binding forces between bacteria and solid surfaces to enable easy removal of the attached bacterial cells, for example by shear forces through salivary flow in the oral cavity [3, 85].

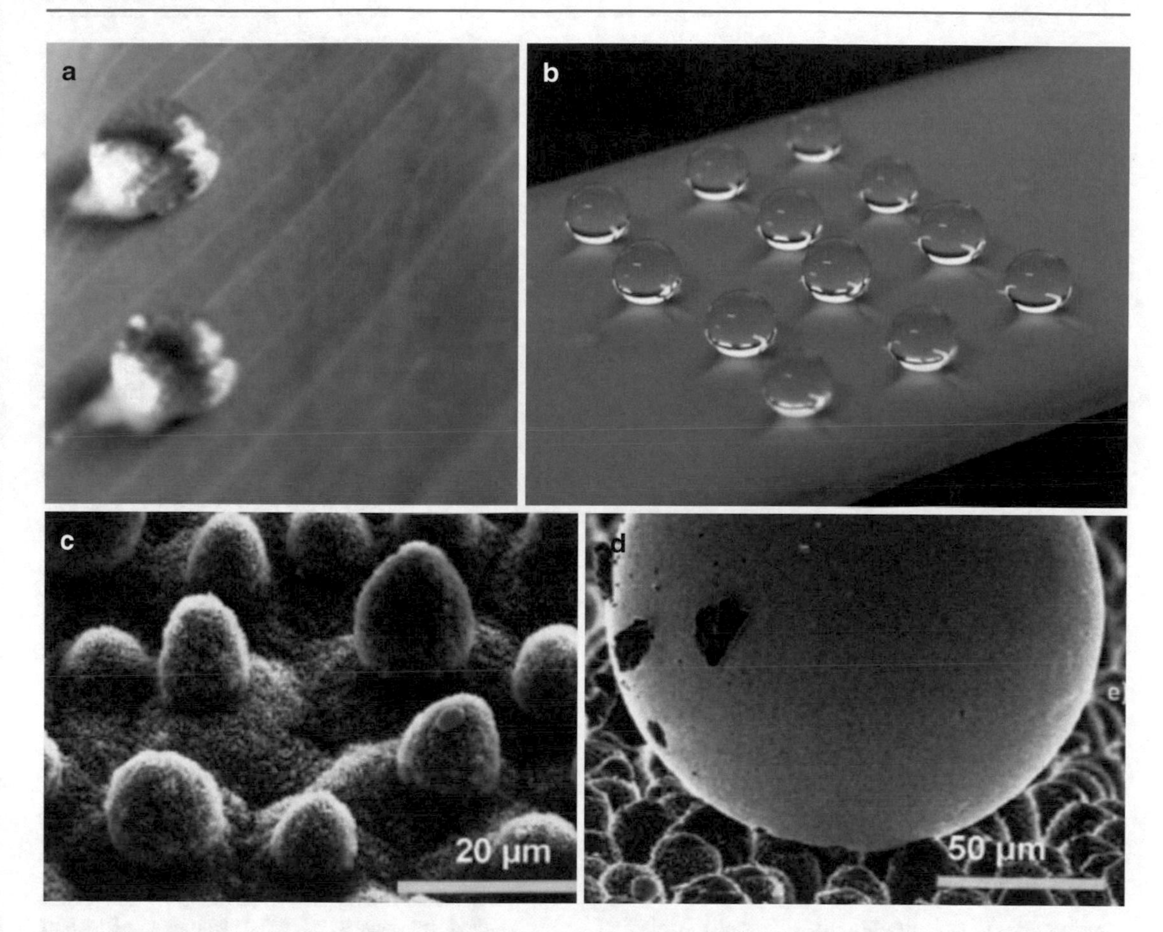

Fig. 5.14 Superhydrophobic surfaces (lotus effect): water droplets on the taro leaf surface (**a**) (*from Y. Yoon et al.: "Hierarchial micro/nano structures for super-hydro-phobic surfaces...", Micro and Nano Systems Letter 2014, licensee SPRINGER*) and on a superhydrophobic coating (**b**) (*from X. Li et al.: "A study on superhydrophobic coating in anti-icing of glass/porcelain insulator", J Sol-Gel Sci Tech 2014, licensee SPRINGER*); scanning electron microscopy of a lotus leaf (*Nelumbo nucifera*) (**c**) and a mercury droplet on the leaf surface (**d**) (***c** and **d** from W. Barthlott et al.: "Purity of the sacred lotus, or escape from contamination in biological surfaces", Planta 1997, licensee SPRINGER*)

5.4 Surface Charge and Zeta Potential

Besides surface roughness and surface free energy, as the predominant factors of influence in biofilm formation on solid substrata, surface charge affects the binding forces between bacteria and abiotic material surfaces, resulting in different bacterial motility, architecture, composition, and physiology of subsequent biofilms [86–90].

5.4.1 Characterization Techniques for Surface Charge

The electrostatic charge density and the electrokinetic potential on bacterial cell surfaces are usually expressed as zeta potential (unit mV) [15]. The zeta potentials of small particles such as microbes cannot be directly measured [15, 91]. Therefore, indirect techniques such as particulate microelectrophoresis have to be applied [15, 91–93]. Hard surface charge may also be represented by the zeta potential and is measured by electrostatic fieldmeter method, electrostatic voltmeter method, electrophoretic mobility, streaming potential measurement, sedimentation potential, or electroosmosis assay [15, 94, 95].

5.4.2 Surface Charge and Biofilm Formation

The bacterial cell wall consists of components such as teichoic acids (linked to either the peptidoglycan or the underlying plasma membrane) or phospholipids and lipopolysaccharides. Through the dissociation of acidic groups such as carboxyl, phosphate, and amino groups (–COOH, $-NH_3$, $-HPO_4$, $-H_2PO_4$, $-HPO_4^-$), most bacteria carry a net negative surface charge at a neutral pH (with very few expectations) [3, 64, 92]. Therefore, a positively charged sample surface is more prone to bacterial adhesion than a negatively charged surface [6, 8, 70, 90, 96]. For example, Rzhepishevska et al. showed less *Pseudomonas aeruginosa* biofilm formation with reduced production of biofilm matrix components (exopolysaccharides) on negatively charged polymer brush surfaces in comparison to samples with positive charge [86]. The salivary pellicle, which rapidly covers all tissues and solid materials in the oral cavity, may equalize surface charges and causes these surfaces to be more negatively charged [8, 97].

There are discrepancies between practical observations and theoretical considerations concerning the influence of surface charge on bacterial adhesion [92]. As surface charge is not the dominant physicochemical property of material substrata, its effect on quantity and quality of in vivo oral biofilm formation should not be overestimated, especially not in the presence of other surface properties and diverse environmental factors [64]. In practice, charge per se may not be sufficient to reduce biofilm formation significantly [8, 97].

References

1. Marsh PD, Moter A, Devine DA. Dental plaque biofilms: communities, conflict and control. Periodontology. 2011;55(1):16–35.
2. Nobbs AH, Lamont RJ, Jenkinson HF. Streptococcus adherence and colonization. Microbiol Mol Biol Rev. 2009;73(3):407–50.
3. Zhang X, Wang L, Levanen E. Superhydrophobic surfaces for the reduction of bacterial adhesion. RSC Adv. 2013;3:12003–20.
4. Quirynen M, Bollen CM. The influence of surface roughness and surface free energy on supra- and subgingival plaque formation in man: a review of the literature. J Clin Periodontol. 1995;22:1–14.
5. Busscher HJ, Norde W, van der Mei HC. Specific molecular recognition and nonspecific contributions to bacterial interaction forces. Appl Environ Microbiol. 2008;74(9):2559–64.
6. Bos R, van der Mei HC, Busscher HJ. Physico-chemistry of initial microbial adhesive interactions—its mechanisms and methods for study. FEMS Microbiol Rev. 1999;23:179–229.
7. Crawford RJ, Webb HK, Truong VK, Hasan J, Ivanova EP. Surface topographical factors influencing bacterial attachment. Adv Colloid Interface Sci. 2012;179-182:142–9.
8. Song F, Koo H, Ren D. Effects of material properties on bacterial adhesion and biofilm formation. J Dent Res. 2015;94(8):1027–34.
9. Antonio PD, Lasalvia M, Perna G, Capozzi V. Scale-independent roughness value of cell membranes studied by means of AFM technique. Biochim Biophys Acta. 2012;1818:3141–8.
10. Boulange-Petermann L, Rault J, Bellon-Fontaine MN. Adhesion of Streptococcus thermophilus to stainless steel with different surface topography and roughness. Biofouling. 1997;11:201–16.
11. Mei L, Busscher HJ, van der Mei HC, Ren Y. Influence of surface roughness on streptococcal adhesion forces to composite resins. Dent Mater. 2011;27:770–8.
12. Hannig M. Transmission electron microscopy of early plaque formation on dental materials in vivo. Eur J Oral Sci. 1999;107:55–64.
13. Rimondini L, Fare S, Brambilla E, Felloni A, Consonni C, Brossa F, Carrassi A. The effect of surface roughness on early in vivo plaque colonization on titanium. J Periodontol. 1997;68:556–62.
14. Scheuermann TR, Camper AK, Hamilton MA. Effects of substratum topography on bacterial adhesion. J Colloid Interface Sci. 1998;208:23–33.
15. Wang Y, Lee SM, Dykes G. The physicochemical process of bacterial attachment to abiotic surfaces: challenges for mechanistic studies, predictability

and the development of control strategies. Crit Rev Microbiol. 2014;41(4):452–64.
16. Kocher T, Langenbeck N, Rosin M, Bernhardt O. Methodology of three-dimensional determination of root surface roughness. J Periodontal Res. 2002;37:125–31.
17. Sander M. A practical guide to the assessment of surface texture. Göttingen: Mahr Feinprüf; 1991.
18. Tschernin M. Oberflächeneigenschaften von Zahnrestaurationsmaterialien, Text. PhD thesis, 2003.
19. Park JB, Yang SM, Ko YK. Evaluation of the surface characteristics of various implant abutment materials using confocal microscopy and white light interferometry. Implant Dent. 2015;24:650–6.
20. Poncin-Epaillard F, Henry J, Marmey P, Legeay G, Debarnot D, Bellon-Fontaine M. Elaboration of highly hydrophobic polymeric surface: a potential strategy to reduce the adhesion of pathogenic bacteria? Mater Sci Eng. 2013;33(3):1152–61.
21. Siegismund D, Undisz A, Germerodt S, Schuster S, Rettenmayr M. Quantification of the interaction between biomaterial surfaces and bacteria by 3-D-modelling. Acta Biomater. 2014;10(1):267–75.
22. Gadelmawla ES, Koura MM, Maksoud TMA, Elewa IM, Soliman HH. Roughness parameters. J Mater Proc Technol. 2002;123:133–45.
23. Surface roughness terminology and parameters. www.predev.com/pdffiles/surface_roughness_terminology_and_parameters.pdf.
24. Rauheitswerte. www.zimob.de/wp-content/uploads/2014/04/Rauheitswerte.pdf.
25. Quick guide to surface roughness measurement. www.mitutoyo.com/wp-content/uploads/2012/11/1984_Surface_Roughness_PG.pdf.
26. Oberflächenbeurteilung. www.ima.uni-stuttgart.de/pdf/studium/bachelor/dt/spezialisierungsfachversuche/HFV_Oberflaechenbeurteilung_2012.pdf.
27. Webb HK, Truong K, Hasan J, Fluke C, Crawford RJ, Ivanova EP. Roughness parameters for standard description of surface nanoarchitecture. Scanning. 2012;34:257–63.
28. Oliveira K, Oliveira T, Teixeira P, et al. Comparison of the adhesion ability of different Salmonella enteritidis serotypes to materials used in kitchens. J Food Prot. 2006;69:2352–6.
29. Ortega MP, Hagiwara T, Watanable H, Sakiyama T, et al. Adhesion behavior and removeability of E. coli on stainless steel surface. Food Control. 1998;21:573–8.
30. An YH, Friedman RJ. Concise review of mechanisms of bacterial adhesion to biomaterial surfaces. J Biomed Mater Res. 1998;43:338–48.
31. Quirynen M, van der Mei HC, Bollen CML. An in vivo study of the influence of the surface roughness of implants on the microbiology of supra- and subgingival plaque. J Dent Res. 1993;72:1304–9.
32. Barnes LM, Lo MF, Adams MR, Chamberlain AHL. Effect of milk proteins on adhesion of bacteria to stainless steel surfaces. Appl Environ Microbiol. 1999;65:4543–8.
33. Faille C, Jullien C, Fontaine F, et al. Adhesion of Bacillus spores and E. coli cells to inert surfaces: role of surface hydrophobicity. Can J Microbiol. 2002;48:728–38.
34. McAllister EW, Carey LC, Brady PG, et al. The role of polymeric surface smoothness of biliary stents in bacterial adherence, biofilm deposition, and stent occlusion. Gastrointest Endosc. 1993;39:422–5.
35. Einwag J, Ulrich A, Gehring F. In-vitro-Plaqueanlagerung an unterschiedliche Füllungsmaterialien. Oralprophylaxe. 1990;12:22–7.
36. Schwartz ML, Phillips RW. Comparison of bacterial accumulation on rough and smooth enamel surfaces. J Periodontal. 1957;28:304–7.
37. Yamauchi M, Yamamoto M. In vitro adherence of microorganisms to denture base resin with different surface texture. Dent Mater J. 1990;9:19–24.
38. Larato DC. Influence of a composite resin restoration on the gingiva. J Prosthet Dent. 1972;28:402–4.
39. Möhrmann W, Regolati B, Renggli HH. Gingival reaction to well-fitted subgingival proximal gold inlays. J Clin Periodontol. 1974;1:120–5.
40. Trivedi SC, Talim ST. The response of human gingiva to restorative materials. J Prosthet Dent. 1973;29:73–80.
41. Weitmann RT, Eames WB. Plaque accumulation on composite surfaces after various finishing procedures. JADA. 1975;91:101–6.
42. Teughels W, Van Assche N, Sliepen I, Quirynen M. Effect of material characteristics and/or surface topography on biofilm development. Clin Oral Implants Res. 2006;17(Suppl 2):68–81.
43. Ionescu A, Wutscher E, Brambilla E, Schneider-Feyrer S, Giessibl FJ, Hahnel S. Influence of surface properties of resin-based composites on in vitro Streptococcus mutans biofilm development. Eur J Oral Sci. 2012;120(5):458–65.
44. Bollen CM, Papaioanno W, Van Eldere J, Schepers E, Quirynen M, van Steenberghe D. The influence of abutment surface roughness on plaque accumulation and peri-implant mucositis. Clin Oral Implants Res. 1996;7:201–11.
45. Quirynen M, Bollen CM, Papaioannou W, Van Eldere J, van Steenberghe D. The influence of titanium abutment surface roughness on plaque accumulation and gingivitis: a short-term observation. Int J Oral Maxillofacial Implants. 1996;11:169–78.
46. Bollen CM, Lambrechts P, Quirynen M. Comparison of surface roughness of oral hard materials to the

threshold surface roughness for bacterial retention: a review of the literature. Dent Mater. 1997;13:258–69.
47. Hannig M. Transmission electron microscopic study of in vivo pellicle formation on dental restorative materials. Eur J Oral Sci. 1997;105(5):422–33.
48. Quirynen M, Marechal M, Busscher HJ, Weerkamp AH, Darius PL, van Steenberghe D. The influence of surface free energy and surface roughness on early plaque formation. J Clin Periodontol. 1990;17:138–44.
49. Tanner J, Robinson C, Soderling E, Vallittu P. Early plaque formation on fibre-reinforced composites in vivo. Clin Oral Invest. 2005;9:154–60.
50. Lafuma A, Quere D. Superhydrophobic states. Nat Mater. 2003;2:457–60.
51. Busscher HJ, Van Pelt AWK, De Boer P, de Jong HP, Arends J. The effect of surface roughening of polymers on measured contact angles of liquids. Colloids Surf. 1984;9:319–31.
52. Owens DK, Wendt RC. Estimation of the surface free energy of polymers. J Appl Polym Sci. 1969;13:1741–7.
53. Young T. An essay on the cohesion of fluids. Philos Trans R Soc Lond. 1805;95:65–87.
54. Tadanaga K, Morinaga J, Matsuda A, Minami T. Superhydrophobic-superhydrophilic micropatterning on flowerlike alumina coating film by the sol-gel method. Chem Mater. 1999;12:590–2.
55. Zhao Q, Liu Y, Abel EW. Effect of temperature on surface free energy of amorphous carbon films. J Colloid Interface Sci. 2004;280:174–83.
56. Bürgers R, Rosentritt M, Handel G. Bacterial adhesion of Streptococcus mutans to provisional fixed prosthodontic material. J Prosthet Dent. 2007;98(6):461–9.
57. Bürgers R, Schneider-Brachert W, Rosentritt M, Handel G, Hahnel S. Candida albicans adhesion to composite resin materials. Clin Oral Invest. 2009;13(3):293–9.
58. Absolom DR, Lamberti FV, Policova Z, Zing W, van Oss CJ, Neumann AW. Surface thermodynamics of bacterial adhesion. Appl Environ Microbiol. 1983;46:90–7.
59. Neumann AW, Good RJ, Hope CJ, Sejpal M. An equation-of-state-approach to determine surface tensions of low-energy solids from contact angles. J Colloid Interface Sci. 1974;49:291–304.
60. Bürgers R, Gerlach T, Hahnel S, Schwarz F, Handel G, Gosau M. In vivo and in vitro biofilm formation on two different titanium implant surfaces. Clin Oral Implants Res. 2010;21:156–64.
61. Kaelble DH. Dispersion-polar surface tension properties of organic solids. J Adhes. 1970;2:66–8.
62. Rabel W. Einige Aspekte der Benetzungstheorie und ihre Anwendung auf die Untersuchung und Veränderung der Oberflächeneigenschaften von Polymeren. Farbe und Lack. 1971;77(10):997–1005.
63. Bürgers R, Cariaga T, Müller R, Rosentritt M, Reischl U, Handel G, Hahnel S. Effects of aging on surface properties and adhesion of Streptococcus mutans on various fissure sealants. Clin Oral Invest. 2009;13:419–26.
64. de Wouters T, Jans C, Niederberger T, Fischer P, Rühs PA. Adhesion potential of intestinal microbes predicted by physico-chemical characterization methods. PLOS One. 2015;10(8):e0136437.
65. Rosenberg M, Gutnick D, Rosenberg E. Adherence of bacteria to hydrocarbons: a simple method for measuring cell-surface hydrophobicity. FEMS Microbiol Lett. 1980;9:29–33.
66. Rosenberg M, Rosenberg E, Judes H, Weiss E. Bacterial adherence to hydrocarbons and to surfaces in the oral cavity. FEMS Microbiol Lett. 1983;20:1–5.
67. Clark WB, Lane MD, Beem E, et al. Relative hydrophobicities of Actinomyces viscosus and Actinomyces naeslundii strains and their adsorption to saliva-treated hydroxyapatite. Infect Immun. 1985;47:730–6.
68. Lindahl M, Faris A, Wadstrom T, Hjerten S. A new test based on "salting out" to measure relative surface hydrophobicity of bacterial cells. Biochim Biophys Acta. 1981;677:471–6.
69. Busscher HJ, Weerkamp AH, van der Mei HV, Van Pelt AWJ, de Jong HP, Arends J. Measurement of the surface free energy of bacterial cell surfaces and its relevance for adhesion. Appl Environ Microbiol. 1984;48:980–3.
70. Morisaki H, Nakagawa K, Shiraishi H. Measurement of attachment force of microbial adhesion. Colloids Surf B Biointerfaces. 1996;6:347–52.
71. Rosenberg M, Kjelleberg S. Hydrophobic interactions: role in bacterial adhesion. In: Marshall KC, editor. Advances in microbial ecology. New York: Plenum; 1986. p. 353–93.
72. Arima Y, Iwata H. Effect of wettability and surface functional groups on protein absorption and cell adhesion using well-defined mixed self-assembled monolayers. Biomaterials. 2007;28:3074–82.
73. Lee JH, Khang G, Lee JW, Lee HB. Interaction of different types of cells of polymer surfaces with wettability gradient. J Colloid Interface Sci. 1998;205:323–30.
74. Li J, Zhang W. Bacterial behaviours on polymer surfaces with organic and inorganic antimicrobial compounds. J Biomed Mater Res A. 2008;88A:448–53.
75. Yang H, Deng Y. Preparation and physical properties of superhydrophobic papers. J Colloid Interface Sci. 2008;325:588–93.
76. Zhao Q, Wang S, Steinhagen HM. Tailored surface free energy of membrane diffusers to minimize microbial adhesion. Appl Surf Sci. 2004;230:371–8.
77. Mabboux F, Ponsonnet L, Morrier JJ, Jaffrezic N, Barsotti O. Surface free energy and bacterial retention to saliva-coated dental implant materials: an in vitro study. Colliods Surf B Biointerfaces. 2004;39(4):1226–31.

78. Hu XL, Ho B, Lim CT, Hsu CS. Thermal treatments modulate bacterial adhesion to dental enamel. J Dent Res. 2011;90(12):1451–6.
79. Busscher HJ, Bos R, van der Mei HC. Initial microbial adhesion is a determinant for the strength of biofilm adhesion. FEMS Microbiol Lett. 1995;128:229–34.
80. Christersson CE, Dunford RG, Glantz PO, Baier RE. Effect of critical surface tension on retention of oral microorganisms. Scand J Dent Res. 1989;97:247–56.
81. van Dijk J, Herkströter F, Busscher H, Weerkamp A, Jansen H, Arends J. Surface free energy and bacterial adhesion. An in vivo study in beagle dogs. J Clin Periodontol. 1987;14:300–4.
82. van Pelt AWJ, De Jong HP, Busscher HJ, Arends J. Dispersion and polar free energies of human enamel. J Biomed Mater Res. 1983;17:637–41.
83. Mabboux F, Ponsonnet L, Morrier JJ, Jaffrezic N, Barsotti O. Surface free energy and bacterial retention to saliva-coated dental implant materials—an in vitro study. Colloids Surf B Biointerfaces. 2004;39:199–205.
84. Bartholett W, Neinhuis C. Purity of the sacred lotus, or escape from contamination in biological surfaces. Planta. 1997;202:1–8.
85. Crick CR, Ismail S, Pratten J, Parkin IP. An investigation into bacterial attachment to n elastomeric superhydrophobic surface prepared via aerosol assisted deposition. Thin Solid Films. 2011;519:3722–7.
86. Rzhepishevska O, Hakobyan S, Ruhal R, Gautrot J, Barbero D, Ramstedt M. The surface charge of antibacterial coatings alters motility and biofilm architecture. Biomater Sci. 2013;1(6):589–602.
87. Terada A, Okuyama K, Nishikawa M, Tsuneda S, Hosomi M. The effect of surface charge property on *Escherichia coli* initial adhesion and subsequent biofilm formation. Biotechnol Bioeng. 2012;109(7):1745–54.
88. Murata H, Koepsel RR, Matyjaszewski K, Russell AJ. Permanent, non-leaching antibacterial surface—2: how high density cationic surfaces kill bacterial cells. Biomaterials. 2007;28(32):4870–9.
89. Cheng G, Zhang Z, Chen S, Bryers JD, Jiang S. Inhibition of bacterial adhesion and biofilm formation on zwitterionic surfaces. Biomaterials. 2007;28(29):4192–9.
90. Gottenbos B, Grijpma DW, van der Mei HC, Feijen J, Busscher HK. Antimicrobial effects of positively charges surfaces on adhering gram-positive and gram-negative bacteria. J Antimicrob Chemother. 2001;48(1):7–13.
91. Wilson WW, Wadeb MM, Holmana SC, Champlinb FR. Status of methods for assessing bacterial cell surface charge properties based on zeta potential measurements. J Microbiol Methods. 2001;43:153–64.
92. Poortinga AT, Bos R, Norde W, Busscher HJ. Electric double layer interactions in bacterial adhesion to surfaces. Surf Sci Rep. 2002;47:1–32.
93. Pederson K. Electrostatic interaction chromatography, a method for assaying the relative surface charges of bacteria. FEMS Microbiol Lett. 1982;12:365–7.
94. Faircloth DC, Allen NL. High resolution measurements of surface charge densities on insulator surfaces. IEEE Trans Dielectr Electr Insul. 2003;10:285–90.
95. Smith WE, Rungis J. Twin adhering conducting spheres in an electric field—an alternative geometry for an electrostatic voltmeter. J Phys E Sci Instrum. 1975;8:379–82.
96. Hermansson M. The DLVO theory in microbial adhesion. Colloids Surf B Biointerface. 1999;14:105–19.
97. Cavalcanti YW, Wilson M, Lewis M, Williams D, Senna PM, Del-Bel-Cury AA, da Silva WJ. Salivary pellicles equalize surfaces' charges and modulate the virulence of Candida albicans biofilm. Arch Oral Biol. 2016;66:129–40.

Complex Polymeric Materials and Their Interaction with Microorganisms

6

Elena Günther, Florian Fuchs, and Sebastian Hahnel

Abstract

In contemporary dentistry, resin-based materials are extensively used for the fabrication of direct and indirect restorations. As the materials available on the market feature increasingly complex chemical compositions and include a variety of ingredients with distinct physical and chemical properties, bioadhesion and biofilm formation on the surface of these materials are difficult to predict. These considerations are particularly relevant for modern resin-based materials that have been tailored with the intention to modulate the formation of biofilms on their surface. The aim of the current summary is to outline the contemporary scientific knowledge regarding the role of complex resin-based materials for bioadhesion and biofilm formation on their surface.

E. Günther · F. Fuchs · S. Hahnel (✉)
Prosthetics and Dental Materials Clinic, Leipzig University, Leipzig, Germany
e-mail: elena.guenther@medizin.uni-leipzig.de; florian.fuchs@medizin.uni-leipzig.de; sebastian.hahnel@medizin.uni-leipzig.de

6.1 Introduction

Over the years, resin-based materials have steadily gained more attention in dentistry, and the materials are still booming. Resin-based materials can be used for the fabrication of both direct and indirect dental restorations, which indicates that the mechanical requirements as well as the physical and chemical properties of these materials must be tailored in dependence on their range of application. Regarding their interaction of resin-based materials with microorganisms and biofilms, the role of these materials has continuously changed. While in the past resin-based materials have regularly been associated with high levels of plaque on their surface, the picture is now less clear and the situation even more complex. Bioadhesion as well as biofilm formation on the surface of dental materials is influenced by numerous effects; moreover, the results of both clinical and laboratory studies on this topic are heavily dependent on the experimental conditions applied. Nevertheless, previous investigations have identified that surface roughness, surface free energy, chemical composition, and surface topography [1, 2] have a significant effect on bioadhesion and biofilm formation on the surface of these materials. A detailed discussion on this topic can be found in Chap. 5. In situ studies are regarded as the gold standard in investigations

A. C. Ionescu, S. Hahnel (eds.), *Oral Biofilms and Modern Dental Materials*,
https://doi.org/10.1007/978-3-030-67388-8_6

dealing with bioadhesion and biofilm formation [3] as they allow biofilm formation under physiological conditions in situ, including the whole spectrum of microorganisms in the oral cavity. However, analysis of bioadhesion and biofilm formation in laboratory trials features the advantages of standardization and high-throughput screening. Moreover, laboratory approaches allow an estimation of bioadhesion and biofilm formation on the surface of experimental materials, which is probably why most investigations were performed under strict laboratory experimental conditions.

To date, numerous different resin-based materials differing in composition and properties are available on the market. The materials can be used either for the fabrication of direct restorations (such as resin-based composites) or for the fabrication of indirect restorations (such as removable dentures), which are usually fabricated in the dental laboratory. Regarding their impact on bioadhesion and biofilm formation in the oral cavity, these materials feature different conditions. Due to the extension of removable denture prostheses, resin-based materials for the fabrication of dentures are in close and extensive contact to gingival tissues. In contrast to resin-based materials designed for the replacement of tooth tissues, they do not have to withstand chewing forces, which coincides with a completely different chemical composition. Thus, with regard to bioadhesion, it is necessary to consider these materials from different point of views.

6.2 Resin-Based Materials for Direct Restorations

6.2.1 Introduction

Contemporary resin-based materials for the fabrication of direct dental restorations are most frequently resin-based composites (RBCs), which include materials that feature a resin matrix supplemented with a sophisticated filler fraction to minimize shrinkage and maximize wear resistance. From a clinical point of view, resin-based composites are distinguished by their consistency, ranging from flowable to condensable materials. However, for estimating the impact of these materials on bioadhesion and biofilm formation, this classification is of little value, and special attention has to be drawn on the chemical composition of these materials. In most current commercial RBC formulations, the resin matrix contains Bis-GMA (bisphenol-A-glycidyl methacrylate), which requires blending with other dimethacrylates such as urethane dimethacrylate (UDMA), triethylene glycol dimethacrylate (TEGDMA), or hydroxyethylmethacrylate (HEMA) due to its high viscosity and its unsuitability to incorporate large filler volume fractions [4]. In order to improve the internal structure of the resin-based composite and to chemically bond the hydrophobic resin matrix to the hydrophilic filler fraction, coupling agents such as silanes are employed. Most of the contemporary resin-based composite materials for direct restorations are photopolymerizable materials including photo initiators, for instance camphorquinone; some formulations have included PPD (acetyl benzoyl/1-phenyl-1,2-propanedione), Lucirin® TPO (monoacetylposphine oxide/2,4,6-trimethylbenzoyldiphenylphosphine oxide), or Irgacure® 819 (bisacylphosphine oxide/phenylbis(2,4,6-trimethylbenzoyl)phosphine oxide) [5]. The filler fraction in modern RBC formulations is complex, and usually comprises nanoscaled filler particles or, in so-called nanohybrid materials, both nano- and microscaled filler particles.

Apart from these classical RBCs, glass ionomer cements can be supplemented with a polymeric ingredient; these materials are usually defined as resin-modified glass ionomer cements (RMGIC). However, these materials are only infrequently used and, in most cases, applied in temporary restorations; thus, the focus of the current outline is set on RBCs.

6.2.2 The Role of RBC Filler Fraction on Bioadhesion and Biofilm Formation

The filler fraction accounts for the surface roughness and topography of RBCs. The relevance of surface roughness for bioadhesion and biofilm formation has extensively been discussed in the last decades. For titanium implant surfaces, Bollen and co-workers have introduced a threshold at 0.2 µm, suggesting that lower values for surface roughness than this threshold do not have an impact on biofilm formation [6]. Similar observations have been published by the Rimondini group, who introduced a threshold at *Ra* of 0.088 μm and *Rz* of 1.027 μm [7]. While it is difficult to simply transfer this threshold to other materials with a more complex composition, the conventional wisdom still is that RBCs foster biofilm formation on their surface in comparison to other tooth-colored dental materials such as ceramics or glass ionomer cements [8]. However, these results appear to be particularly true for early RBC formulations, which included large filler particles and featured only insufficient chemical bonding between resin matrix and filler fraction. These materials were difficult to polish, and as a result from hydrolytic effects and wear, the disintegration of fillers from the RBC surfaces continuously produced surfaces with high surface roughness. As microorganisms preferentially adhere to surface imperfections that provide shelter from shear stresses, this phenomenon fosters the adhesion of microorganisms. However, coinciding with a continuous improvement of the mechanical properties of RBCs, materials with very tiny filler particles have been developed and the problem of insufficient bonding between filler particles and resin matrix has largely been overcome.

Figure 6.1 illustrates the history development of filler fraction in resin-based composites. Different shapes and particle size distributions of fillers of modern resin-based composite materials were determined by scanning electron microscopy and displayed in Figure 6.2. Early investigations showed an accumulation of pellicle on filler particles and the crevice between

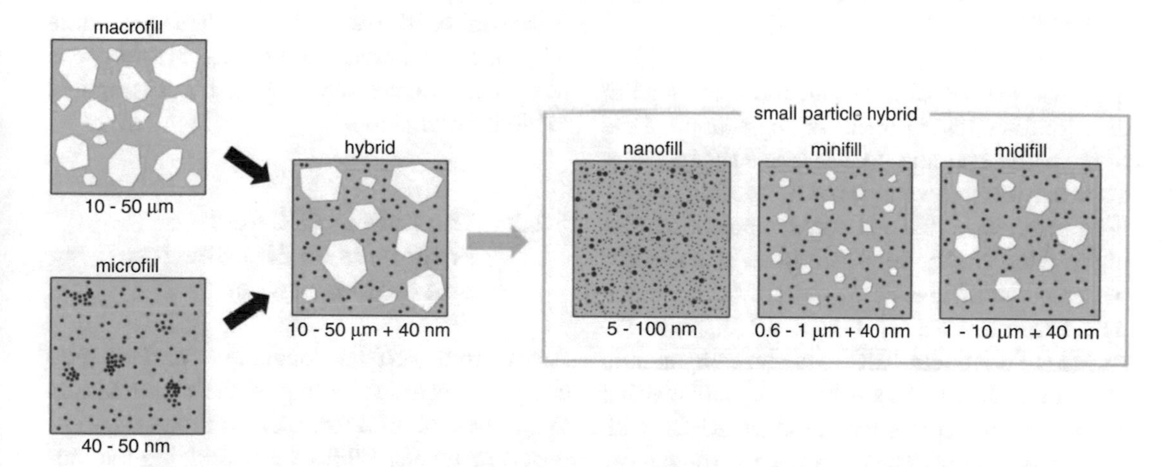

Fig. 6.1 The development of the state of the art of dental composite formulations based on filler particle modifications, based on Ferracane [5], p. 32

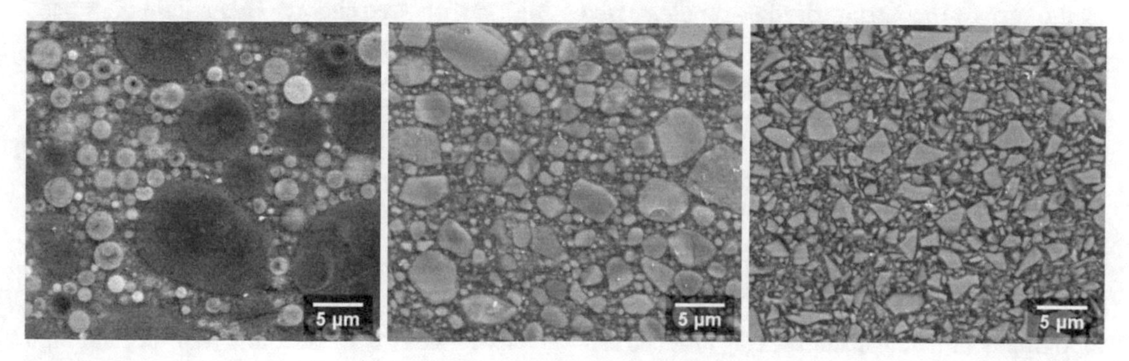

Fig. 6.2 Scanning electron microscopy pictures (secondary electron imaging) of inorganic fillers in conventional resin-based composite material

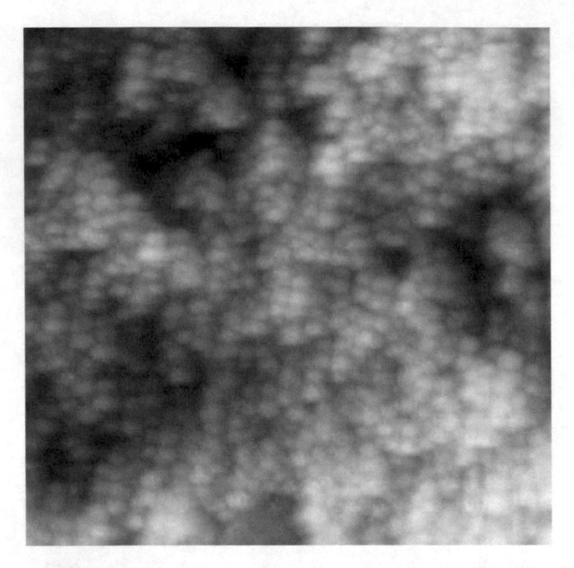

Fig. 6.3 Atomic force microscopic image of the surface of a modern resin-based composite material. (E. Wutscher, Institute of Experimental and Applied Physics, University of Regensburg)

filler and matrix. This circumstance can lead to deterioration, loss of fillers, or degradation of the filler-matrix bonding [9]. Modern RBC formulations may feature volume fractions of up to 89% filler particles [10], and recent clinical studies highlighted that the surface roughness of modern materials is far lower than the previously introduced thresholds [11]. Figure 6.3 displays the surface of a modern RBC as analyzed by atomic force microscopy. As a result, it is unlikely that further modifications will produce RBCs with relevantly diminished surface roughness. However, apart from pure surface roughness some laboratory studies have underlined that biofilm formation can be significantly impacted by the surface topography of filler-supplemented resin-based materials. Data from various groups have demonstrated that different polishing regimes can significantly impact biofilm formation on the surface of a single material, although the differences in surface roughness between these interfaces were negligible [11–13]. These results respond to the knowledge that has been gathered in other biological systems and underline that effective modification of the surface topography of dental materials may produce surfaces with antifouling properties [1]. For instance, the topography of shark skin is less susceptible for bioadhesion and serves as an inspiration for several micropatterned surfaces developed for the use in other fields of medicine as well as marine environments [14–16]. Surface structuring leads to increased water contact angles and therewith hydrophobicity. Furthermore, micro-structured surface patterns trap air, which decreases the available contact area between the substratum surface and microorganisms. Also, quorum sensing between microorganisms seems to be reduced due to topographical barriers. Frenzel and co-workers developed an approach to produce different composite surface structures to reduce bioadhesion on dental restorations. One conceivable future implementation of the advantages of biomaterials' surfaces is engineering matrix strips which microstructure the composite during the placement of the restoration [1]. This procedure might allow the production of direct restorations with optimized surfaces that are less susceptible to biofilm formation.

6.2.3 The Role of RBC Surface Properties on Bioadhesion and Biofilm Formation

Apart from surface roughness, surface free energy is regarded as one of the major factors that impact biofilm formation on the surface of dental materials. While the underlying thermodynamical principles are complex, a simple rule of thumb is that surfaces with low surface free energy attract less plaque than surfaces with high surface free energy. This relation has been proven for a variety of simple polymeric materials such as polytetrafluorethylene (PFTE) or polyethylene (PE) in several in situ experiments [2]; however, the relation between surface free energy and biofilm formation is more complex and less clear for complex materials such as RBCs. RBCs consist of chemically distinct ingredients with different surface free energies, which finally results in a complex surface that includes areas with both low (matrix) and high (fillers) surface free energy (cf. Fig. 6.1). Thus, relations between surface free energy and microbial adhesion and subsequent biofilm formation are hard to establish, which might serve as an explanation why conflicting results have been reported regarding the role of the surface free energy of a RBC on biofilm formation [17, 18]. In some laboratory studies, biofilm formation was lower on the surface of a RBC material with a distinct hydrophobic resin matrix based on siloranes than on the surface of conventionally applied methacrylate-based materials [19, 20]. However, in situ studies did not support the results of the laboratory investigations [21], which underlines that the results from laboratory approaches cannot be simply transferred into clinical settings. As the proportion of the resin matrix on the surface of a RBC is low due to the abundancy of the filler fraction, it is likely that differences in surface free energy resulting from a variation of the resin matrix may only affect bioadhesion and biofilm formation under controlled and very strict experimental conditions. Judging from the current evidence, it appears that surface free energy cannot serve as a reliable predictor of bioadhesion and biofilm formation on the surface of complex RBCs. Overall, the importance of RBC surface properties on bioadhesion seems to decrease with increasing biofilm formation time and growing biofilm layer [22].

6.2.4 The Role of the RBC Resin Matrix on Bioadhesion and Biofilm Formation

In comparison to filler fraction, only little effort has so far been made to identify and elucidate the interaction of the resin matrix with bioadhesion. While it is clear from the history of the filler fraction in RBCs that the proportion of the resin matrix in the surface of RBCs has gradually diminished, some researchers have highlighted that the resin matrix has a relevant impact on biofilm formation on the surface of RBCs [11]. Brambilla and co-workers have shown that *Streptococcus mutans* biofilm formation decreases with increasing curing time, which has been attributed to a decreased concentration of unpolymerized monomers [23]. Previous studies have also highlighted that the colonization of specimens fabricated from different experimental resins with *Streptococcus mutans* is different despite similar surface properties [24], which underlines the existence of an effect of the resin matrix on bioadhesion. However, the exact mechanisms which are responsible for these results are not yet clear. It is well known that polymerization of resin-based materials is never complete, and modern mixtures feature a degree of conversion ranging around 60–70%. Leakage of unpolymerized resin monomers as well as biodegradation of the resin matrix by oral microorganisms might have an impact on oral microorganisms [11]. In the past, it has been suggested that monomers such as UDMA, EGDMA, DEGDMA, and TEGDMA may promote growth and proliferation of cariogenic microorganisms [25, 26]; however, recent publications could not corroborate this hypothesis [27].

6.2.5 Modifications of RBCs with Antimicrobial Agents

Several approaches to equip RBCs with antimicrobial properties have been described, including the incorporation of agents like silver ions [28, 29], zinc oxide nanoparticles [30], quaternary ammonium derivatives [31], chlorhexidine acetate [32], and many others. Antimicrobial agents delay, reduce, or avoid biofilm formation through direct contact or leaching. However, these agents may lead to impaired mechanical properties and decreased degrees of conversion. Also, the antibacterial effects are often temporary [33], featuring a burst effect followed by a rapid decrease. Early studies indicated an antimicrobial effect on *Streptococcus mutans* by quaternary ammonium polyethylenimine nanoparticles for at least 1 month [31] and silver-supplemented materials for 6 months [34]. At the same time, little or no release of silver or quaternary ammonium was observed. Moreover, Yoshida and co-workers reported that supplementing RBCs with different silver agents produced materials with very distinct mechanical properties [34], which underlines that it is not indifferent to the antimicrobial agent used. With regard to this aspect, it has been reported that quaternary ammonium polyethylenimine nanoparticles do not compromise the mechanical properties of a supplemented RBC [31]. Other studies suggest even higher mechanical strength (diametral tensile strength, fracture toughness) in RBCs supplemented with titanium and silver-tin-copper filler particles [35]. More information on this can be found in Chaps. 8, 9, and 10.

6.2.6 The Role of Degradation of RBCs by Microorganisms: A Circulus Vitiosus?

However, although no simple relations between the composition and availability of resin monomers and bioadhesion have yet been identified, it is undoubted that oral microorganisms interact with the matrix constituents of RBCs. It has been highlighted that the presence of biofilms on the surface of RBCs leads to deterioration of the RBC surface [36–38]. This phenomenon is due to biodegradation of the resin matrix, an effect that has frequently been addressed in dental materials science and is caused by esterases from cariogenic bacteria [39]. As acids may also provoke degradation of RBCs [40], it is likely that acids produced by cariogenic microorganisms can also affect the surface of RBCs. Recent literature indicates that deterioration of RBC surfaces by microorganisms is dependent on the bacterial strain as well as the composition of the resin matrix. Bis-GMA-free formulations did not show changes in surface roughness after exposition to cariogenic streptococci, while Bis-GMA-containing RBC formulations were significantly affected [41]. Results from in situ studies indicate that RBCs with urethane dimethacrylate matrix are less vulnerable against deterioration than RBCs with mixed matrices (UDMA, Bis-GMA, and DDMA). Moreover, an accumulation of pellicle was observed on filler particles and between filler and the matrix [9]. Simple monospecies biofilms including cariogenic microorganisms such as *Streptococcus mutans* seem to have a more distinct effect on the surface than multispecies biofilms [41]. While recent studies identified only slight increases in surface roughness in a nanometer range after exposition to the various biofilms [41], early studies showed that the surface of RBCs is relevantly affected [37, 39]. Thus, it is not yet clear as to how far biodeterioration of RBCs by microorganisms produces an impaired RBC surface that substantially fosters subsequent bioadhesion. While current scientific evidence is still scarce, it might be recommended to regularly polish RBC restorations in order to minimize a potential negative effect of a roughened surface.

6.3 Resin-Based Materials for Indirect Restorations

6.3.1 Introduction

Resin-based materials for indirect dental restorations include materials for the fabrication of removable dentures as well as polymer-based materials for the CAD/CAM fabrication of indirect restorations such as inlays, partial crowns, crowns, and fixed-partial dentures.

Resin-based materials have been used for decades for the fabrication of removable dental prostheses. For these applications, poly(methyl methacrylate) (PMMA) is still regarded as the material of choice, although some other materials have been introduced as well. Alternative denture materials can be divided into methacrylate-containing and methacrylate-free materials (cf. Table 6.1/Fig. 6.4). The latter group features the

Table 6.1 Examples for methacrylate-containing and methacrylate-free materials

Methyl methacrylate-containing materials	Methyl methacrylate-free materials
Thermoplastic poly(methyl methacrylate), *PMMA*	Polyoxymethylene, *POM*
	Polyamide, *PA* (Valplast®)
Vinyl polymer (vinyl chloride and vinyl acetate, contains *MMA* as copolymer, Luxene®)	Polyurethane dimethacrylate (light curing, Eclipse®)
	Poly(aryl ether ketones), *PAEK*
	Poly(ether ether ketone), *PEEK*
	Poly(ether ketone ketone), *PEKK*

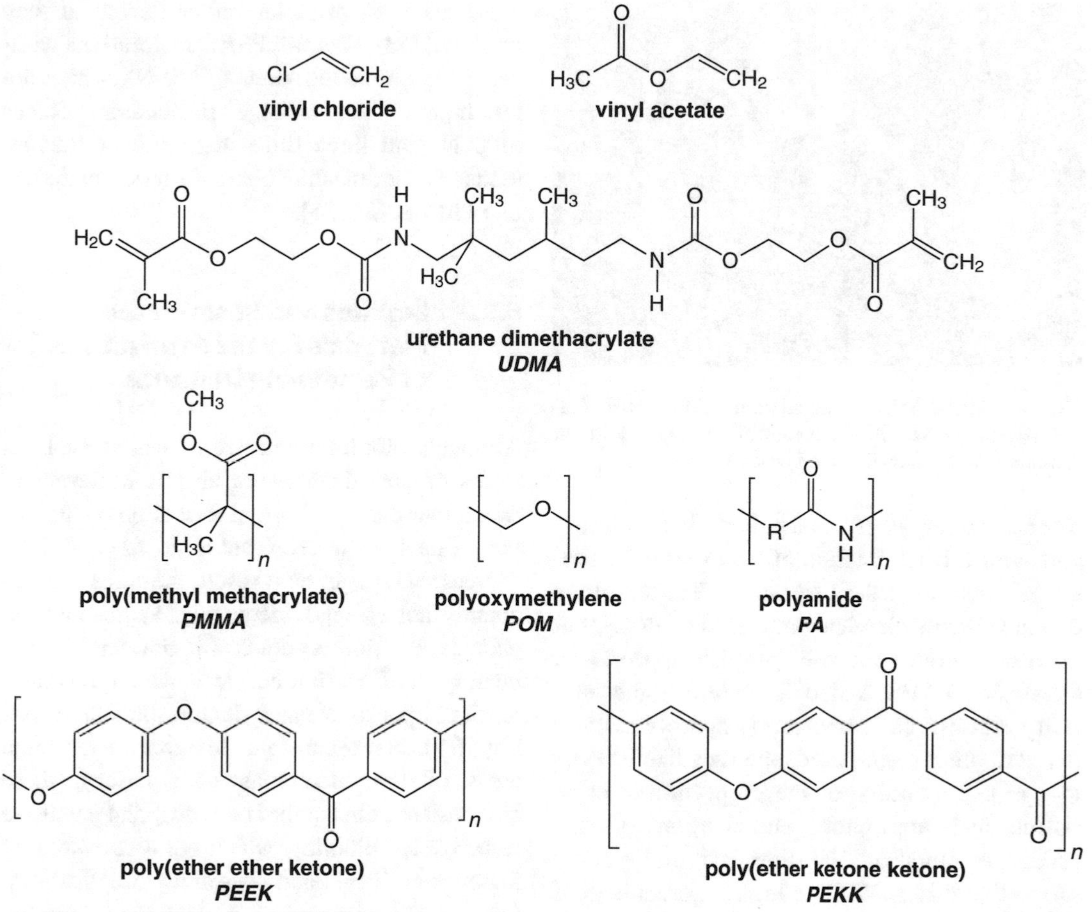

Fig. 6.4 Chemical structure of monomers and polymers of modern resin-based materials

advantages of little elution of monomers and high biocompatibility. Thus, they are particularly suitable in patients with allergies. At the same time, these materials are difficult to repair and to reline since they are chemically inert [42]. Some of the materials are very flexible and can be used in patients with microstomia (e.g., polyamide).

Further innovations in this field include machinable PMMA-based materials for CAD/CAM fabrication of denture bases. Moreover, innovative materials from the family of poly(aryl ether ketone) (PAEK) such as poly(ether ether ketone) (PEEK) or poly(ether ketone ketone) (PEKK) have been introduced in the last years. In contrast to PMMA-based dentures, these polymers feature the advantages of CAD/CAM fabrication, improved mechanical properties [43], low allergenic potential [44], as well as low weight [45]. In case of PEEK or PEKK, even complex tooth- or implant-supported denture prostheses can be fabricated from the polymeric material without a supporting alloy framework (cf. Fig. 6.5) [45, 46]. PEEK and PEKK differ in their ratio of ketone and ether groups. As PEKK contains more ketone groups it is slightly stiffer than PEEK. Its mechanical, optical, and chemical properties are similar to PEEK [47]. Since PEEK was introduced earlier into the dental market than PEKK and is available for different fabrication techniques (e.g., heat pressing, CAD/CAM) it is more popular than PEKK. Therefore, studies on PEKK are still scarce. While PEKK is supplemented with a filler fraction of about 10 wt% titanium dioxide

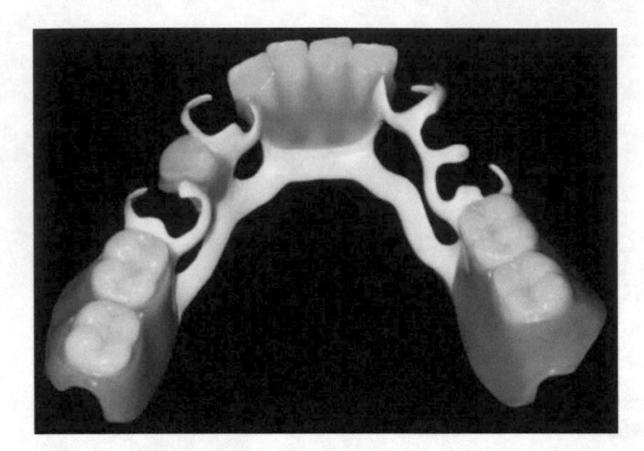

Fig. 6.5 CAD/CAM-fabricated removable partial denture prosthesis with PEEK framework. (Image by Ingolf Riemer, Universitätsklinikum Leipzig)

filler particles, PEEK is available both without and with a filler fraction of up to 30 wt% titanium dioxide filler particles. These fillers enhance the mechanical properties of PAEK but do also account for their grayish appearance. Generally, PAEK materials feature higher stability, rigidity, and resistance to hydrolysis compared to other resin-based materials like PMMA [48, 49]. Depending on the proportion of crystalline and amorphous contributions to the PAEK formulation, its chemical and optical properties differ: While a higher percentage of crystalline parts promotes resistance to acids, alkalis, and organic solvents, a higher proportion of the amorphous phase coincides with enhanced translucency [50]. In order to combine good chemical and aesthetic properties, PAEK restorations with a higher percentage of crystalline parts can be veneered or lined using resin-based materials [46, 51]. Crystalline PEKK, which shows higher flexural and tensile strength, is preferably used for the fabrication of crowns and fixed dental prosthesis, while amorphous PEKK can be applied for the fabrication of removable prosthesis [52]. Stawarczyk and co-workers identified increased fracture loads for milled PEEK restorations compared to those which had been pressed from granular. Materials which were pressed from industrially fabricated pellets and milled PEEK restorations showed spontaneous and brittle fractures in the pontic areas without deformation, whereas pressed materials from granular rather deformed than fractured [53]. Overall, PEEK restorations withstand high breaking loads (1300 N), which are far higher than average masticatory forces (400 N) and three times higher than fracture loads of other machinable resin-based materials like PMMA [54, 55].

6.3.2 Biofilms and Resin-Based Materials for the Fabrication of Removable Dentures

Although biofilm-associated diseases such as caries or periodontitis are almost irrelevant in edentulous patients, biofilm formation on the surface of dentures is a relevant issue. Manual skills decrease with age; as a result, dentures are frequently not cleaned adequately. In hospitalized patients, the time available for oral care by the nursing staff is limited, too, which regularly results in poorly cleaned denture prostheses (cf. Fig. 6.7). As removable dentures cover large areas of the edentulous gum tissues, these circumstances make them an ideal and extensive reservoir for biofilms, which are a relevant risk factor for local and systemic implications. Denture-related stomatitis constitutes a common local biofilm-associated disease in denture wearers with a prevalence of 15% to over 70% [56]. *Candida albicans* plays a major role in the pathogenesis of denture-related stomatitis [57]. This fungus has three morphological forms: blastospores, hyphae, and pseudohyphae. The morphological transformation of *C. albicans* from blastospores to hyphae coincides with a maturing process of the biofilm and induces an increased pathogenicity and virulence [57–59]. This transformation seems to be regulated by secreted proteases and their activity [60]. Aspartate proteinases are among the most frequently discussed virulence factors of *C. albicans*, as they contribute decisively to the degradation of host proteins and promote the invasion of the fungus into the oral mucous membranes as well as the development of Candida-associated prosthetic stomatitis [61]. *C. albicans* cells organized in biofilms secrete higher levels of aspartate proteases than planktonic cells [62]. Moreover, the activity of the proteinases correlates with the severity of denture stomatitis [60]. Since the risk of tissue invasion of *C. albicans* increases with the presence of fungal hyphae in mature biofilms, regular oral and denture hygiene is essential [59, 63]. Besides oral hygiene, surface properties as well as surface topography of dental materials shall be optimized in order to minimize fungal and microbial adherence. The substratum surface properties may promote a genetic response which leads to the transformation from blastospores to hyphae [64–66]. Apart from that, high polar contribution to surface free energy increases the proliferation of *C. albicans*, e.g., on urethane dimethacrylate (UDMA)-based denture base materials and soft denture liners (siloxane based) [67]. *C. albicans* hyphae seem to adhere preferentially to hydrophobic rather than hydrophilic surfaces [64, 68]. Moreover, porosities in the denture base foster microbial and fungal adherence by increasing the available surface [69]. An increased number of hyphae was observed in biofilms on siloxane-based soft denture liners compared to PMMA- and UDMA-based denture base materials [59]. Fungal proliferation promotes the deterioration of the surface of denture liners, which may coincide with further irritations of the mucosa [66]. Therefore, long-term use of soft denture liners cannot be recommended. Moreover, denture age and continuous denture wearing are important factors in the development of denture-related stomatitis [57]. Figure 6.6 illustrates differences in surface appearance between new and aged denture bases.

For the relining of denture prostheses, materials based on acrylate or silicone (soft denture liner) can be used. The physical properties of relining materials seem to affect biofilm formation significantly [59, 70]: For instance, *C. albicans* preferentially adheres to hydrophobic rather than hydrophilic surfaces [64]. Hence, hydrophilic coatings of denture surfaces might decrease the attachment of hydrophobic fungal cells like those of *C. albicans* [68]. Further investigations showed that *C. albicans* adherence on polyamides is higher than on PMMA [71]. Materials with smooth surfaces as well as low surface free energy feature less fungal-microbial adherence than materials with rough surfaces and high surface free energy [22]. Surface properties seem to influence especially the early fungal-microbial colonization on dental materials; with maturation of the biofilms, the differences in biofilm formation between various materials gradually diminish [22].

Apart from biofilm-induced diseases such as denture-related stomatitis biofilms on removable denture prostheses may also have other systemic consequences. As respiratory pathogenic microorganisms have been identified in denture plaque [72, 73], denture wearing has been associated with the occurrence of pneumonias in elderly patients [74, 75]. Pneumonia is a common infection in elderly people and constitutes the most frequent cause of mortality from nosocomial infection in elderly patients with a mortality rate

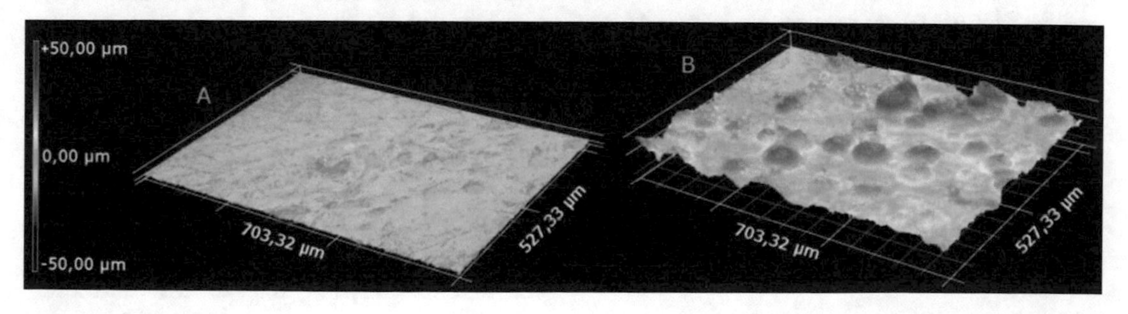

Fig. 6.6 Simulated surface of a new (**A**) and a 30-year-old (**B**) denture prosthesis based on microscopic evaluation

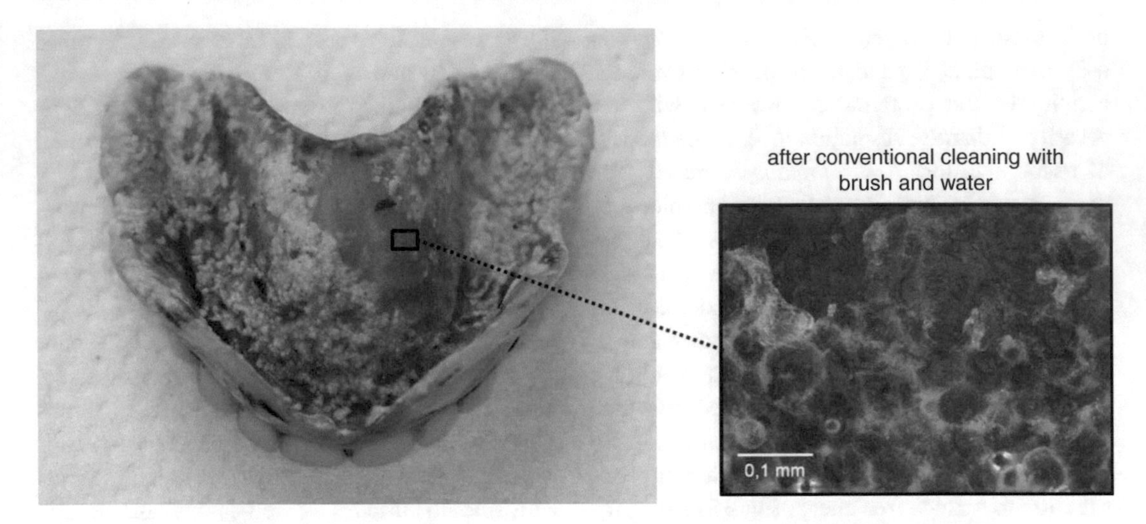

Fig. 6.7 Removable dental prosthesis with extended accumulation of plaque, caused by neglected hygiene by a hospitalized 90-year-old woman

up to 25% [76, 77]. Aspiration of oropharyngeal bacteria into the lungs due to dysphagia and nocturnal denture wearing as well as weakened host defense mechanisms may lead to respiratory infections [78, 79]. Especially for hospitalized patients it is often difficult to maintain a sufficient oral care due to impaired cleaning abilities or limited help provided by the nursing staff (cf. Fig. 6.7) [74]. Moreover, elderly patients have difficulties in accessing professional dental care and consequently appear at the dental office when having denture problems or pain [74, 79]. Thus, regular oral care is required [72, 77, 80], which includes the mechanical removal of denture plaque. Figure 6.7 displays a removable denture prosthesis with extensive accumulation of biofilms. Several studies showed that oral hygiene may have a positive effect on morbidity and mortality from pneumonia: One of ten deaths due to pneumonia in nursing homes may be prevented by improving oral hygiene [77]. Another approach would be the development of materials which feature as little bioadhesion as possible. However, with regard to respiratory microorganisms no scientific data are currently available regarding an impact of the substratum material.

6.3.3 Biofilms on Resin-Based Materials for Fixed Dental Restorations

Scientific data on the formation of biofilms on these materials is limited, as most of the materials have only recently been introduced into the dental market. Nevertheless, it might be possible that resin-based materials polymerized in an industrial setting feature reduced biofilm formation in comparison to their counterparts which are polymerized under clinical conditions. It is conceivable that milled restorations and restorations made by heat pressing from industrially fabricated pellets might show lower biofilm formation due to higher homogeneity as they also feature better mechanical properties compared to those of non-industrially fabricated materials [53]. A recent study contrasted several restorative materials for CAD/CAM fabrication regarding biofilm formation. Interestingly, the acrylate-based material featured significantly lower biofilm formation in comparison to the polymer-infiltrated ceramic as well as zirconia [81]. The authors assumed that materials with a higher ratio of organic constituents (polymer) feature less biofilm formation than materials with a higher ratio of inorganic constituents (ceramic). This conclusion is surprising as several researchers have reported conflicting results [82–84]. However, some studies also confirmed lower biofilm adhesion on composite than, e.g., on ceramic surfaces [85]. Certainly, these findings depend on the used type of ceramic or polymer, its finishing, the used bacteria species, and, finally, the study design. Hence, these results may not be easily transferred into clinical considerations, underlining the relevance of well-designed and adequate in vivo studies.

References

1. Frenzel N, Maenz S, Sanz Beltrán V, Völpel A, Heyder M, Sigusch BW, et al. Template assisted surface microstructuring of flowable dental composites and its effect on microbial adhesion properties. Dent Mater. 2016;32(3):476–87. https://doi.org/10.1016/j.dental.2015.12.016.
2. Teughels W, van Assche N, Sliepen I, Quirynen M. Effect of material characteristics and/or surface topography on biofilm development. Clin Oral Implants Res. 2006;17(Suppl 2):68–81. https://doi.org/10.1111/j.1600-0501.2006.01353.x.
3. Hannig C, Hannig M. The oral cavity—a key system to understand substratum-dependent bioadhesion on solid surfaces in man. Clin Oral Investig. 2009;13(2):123–39. https://doi.org/10.1007/s00784-008-0243-3.
4. Hahnel S, Dowling AH, El-Safty S, Fleming GJP. The influence of monomeric resin and filler characteristics on the performance of experimental resin-based composites (RBCs) derived from a commercial formulation. Dent Mater. 2012;28(4):416–23. https://doi.org/10.1016/j.dental.2011.11.016.
5. Ferracane JL. Resin composite—state of the art. Dent Mater. 2011;27(1):29–38. https://doi.org/10.1016/j.dental.2010.10.020.
6. Bollen CM, Lambrechts P, Quirynen M. Comparison of surface roughness of oral hard materials to the threshold surface roughness for bacterial plaque retention: a review of the literature. Dent Mater. 1997;13(4):258–69.
7. Rimondini L, Farè S, Brambilla E, Felloni A, Consonni C, Brossa F, Carrassi A. The effect of surface roughness on early in vivo plaque colonization on titanium. J Periodontol. 1997;68(6):556–62. https://doi.org/10.1902/jop.1997.68.6.556.
8. Svanberg M, Mjör IA, Orstavik D. Mutans streptococci in plaque from margins of amalgam, composite, and glass-ionomer restorations. J Dent Res. 1990;69(3):861–4. https://doi.org/10.1177/00220345900690030601.
9. Gröger G, Rosentritt M, Behr M, Schröder J, Handel G. Dental resin materials in vivo—TEM results after one year: a pilot study. J Mater Sci Mater Med. 2006;17(9):825–8. https://doi.org/10.1007/s10856-006-9841-2.
10. Wang R, Habib E, Zhu XX. Evaluation of the filler packing structures in dental resin composites: From theory to practice. Dent Mater. 2018;34(7):1014–23. https://doi.org/10.1016/j.dental.2018.03.022.
11. Ionescu A, Wutscher E, Brambilla E, Schneider-Feyrer S, Giessibl FJ, Hahnel S. Influence of surface properties of resin-based composites on in vitro Streptococcus mutans biofilm development. Eur J Oral Sci. 2012;120(5):458–65. https://doi.org/10.1111/j.1600-0722.2012.00983.x.
12. Hahnel S, Wastl DS, Schneider-Feyrer S, Giessibl FJ, Brambilla E, Cazzaniga G, Ionescu A. Streptococcus mutans biofilm formation and release of fluoride from experimental resin-based composites depending on surface treatment and S-PRG filler particle fraction. J Adhes Dent. 2014;16(4):313–21. https://doi.org/10.3290/j.jad.a31800.
13. Park JW, Song CW, Jung JH, Ahn SJ, Ferracane JL. The effects of surface roughness of composite resin on biofilm formation of Streptococcus mutans in the presence of saliva. Oper Dent. 2012;37(5):532–9. https://doi.org/10.2341/11-371-L.
14. Mann EE, Manna D, Mettetal MR, May RM, Dannemiller EM, Chung KK, et al. Surface micropattern limits bacterial contamination. Antimicrob Resist Infect Control. 2014;3:28. https://doi.org/10.1186/2047-2994-3-28.
15. Reddy ST, Chung KK, McDaniel CJ, Darouiche RO, Landman J, Brennan AB. Micropatterned surfaces for reducing the risk of catheter-associated urinary tract infection: an in vitro study on the effect of sharklet micropatterned surfaces to inhibit bacterial colonization and migration of uropathogenic Escherichia coli. J Endourol. 2011;25(9):1547–52. https://doi.org/10.1089/end.2010.0611.
16. Schumacher JF, Long CJ, Callow ME, Finlay JA, Callow JA, Brennan AB. Engineered nanoforce gradients for inhibition of settlement (attachment) of swimming algal spores. Langmuir. 2008;24(9):4931–7. https://doi.org/10.1021/la703421v.
17. Hahnel S, Ionescu AC, Cazzaniga G, Ottobelli M, Brambilla E. Biofilm formation and release of fluoride from dental restorative materials in relation to their surface properties. J Dent. 2017;60:14–24.
18. Ionescu AC, Hahnel S, Cazzaniga G, Ottobelli M, Braga RR, Rodrigues MC, Brambilla E. Streptococcus mutans adherence and biofilm formation on experimental composites containing dicalcium phosphate dihydrate nanoparticles. J Mater Sci Mater Med. 2017;28(7):108.
19. Brambilla E, Ionescu A, Cazzaniga G, Ottobelli M. Influence of light-curing parameters on biofilm

development and flexural strength of a silorane-based composite. Oper Dent. 2016;41(2):219–27. https://doi.org/10.2341/14-279-L.

20. Buergers R, Schneider-Brachert W, Hahnel S, Rosentritt M, Handel G. Streptococcal adhesion to novel low-shrink silorane-based restorative. Dent Mater. 2009;25(2):269–75. https://doi.org/10.1016/j.dental.2008.07.011.
21. Claro-Pereira D, Sampaio-Maia B, Ferreira C, Rodrigues A, Melo LF, Vasconcelos MR. In situ evaluation of a new silorane-based composite resin's bioadhesion properties. Dent Mater. 2011;27(12):1238–45. https://doi.org/10.1016/j.dental.2011.08.401.
22. Hahnel S, Wieser A, Lang R, Rosentritt M. Biofilm formation on the surface of modern implant abutment materials. Clin Oral Implants Res. 2015;26(11):1297–301. https://doi.org/10.1111/clr.12454.
23. Brambilla E, Gagliani M, Ionescu A, Fadini L, García-Godoy F. The influence of light-curing time on the bacterial colonization of resin composite surfaces. Dent Mater. 2009;25(9):1067–72. https://doi.org/10.1016/j.dental.2009.02.012.
24. Hahnel S, Rosentritt M, Buergers R, Handel G. Surface properties and in vitro Streptococcus mutans adhesion to dental resin polymers. J Mater Sci Mater Med. 2008;19(7):2619–27. https://doi.org/10.1007/s10856-007-3352-7.
25. Hansel C, Leyhausen G, Mai UE, Geurtsen W. Effects of various resin composite (co)monomers and extracts on two caries-associated micro-organisms in vitro. J Dent Res. 1998;77(1):60–7. https://doi.org/10.1177/00220345980770010601.
26. Kawai K, Torii M, Tuschitani Y. Effect of resin components on the growth of Streptococcus mutans. J Osaka Univ Dent Sch. 1988;28:161–70.
27. Nedeljkovic I, Yoshihara K, de Munck J, Teughels W, van Meerbeek B, van Landuyt KL. No evidence for the growth-stimulating effect of monomers on cariogenic Streptococci. Clin Oral Investig. 2017;21(5):1861–9. https://doi.org/10.1007/s00784-016-1972-3.
28. Buergers R, Eidt A, Frankenberger R, Rosentritt M, Schweikl H, Handel G, Hahnel S. The anti-adherence activity and bactericidal effect of microparticulate silver additives in composite resin materials. Arch Oral Biol. 2009;54(6):595–601. https://doi.org/10.1016/j.archoralbio.2009.03.004.
29. Yamamoto K, Ohashi S, Aono M, Kokubo T, Yamada I, Yamauchi J. Antibacterial activity of silver ions implanted in SiO2 filler on oral streptococci. Dent Mater. 1996;12(4):227–9. https://doi.org/10.1016/S0109-5641(96)80027-3.
30. Tavassoli Hojati S, Alaghemand H, Hamze F, Ahmadian Babaki F, Rajab-Nia R, Rezvani MB, et al. Antibacterial, physical and mechanical properties of flowable resin composites containing zinc oxide nanoparticles. Dent Mater. 2013;29(5):495–505. https://doi.org/10.1016/j.dental.2013.03.011.
31. Beyth N, Yudovin-Farber I, Bahir R, Domb AJ, Weiss EI. Antibacterial activity of dental composites containing quaternary ammonium polyethylenimine nanoparticles against Streptococcus mutans. Biomaterials. 2006;27(21):3995–4002. https://doi.org/10.1016/j.biomaterials.2006.03.003.
32. Leung D, Spratt DA, Pratten J, Gulabivala K, Mordan NJ, Young AM. Chlorhexidine-releasing methacrylate dental composite materials. Biomaterials. 2005;26(34):7145–53. https://doi.org/10.1016/j.biomaterials.2005.05.014.
33. Jandt KD, Sigusch BW. Future perspectives of resin-based dental materials. Dent Mater. 2009;25(8):1001–6. https://doi.org/10.1016/j.dental.2009.02.009.
34. Yoshida K, Tanagawa M, Atsuta M. Characterization and inhibitory effect of antibacterial dental resin composites incorporating silver-supported materials. J Biomed Mater Res. 1999;47(4):516–22.
35. Jandt KD, Al-Jasser AMO, Al-Ateeq K, Vowles RW, Allen GC. Mechanical properties and radiopacity of experimental glass-silica-metal hybrid composites. Dent Mater. 2002;18(6):429–35.
36. Beyth N, Bahir R, Matalon S, Domb AJ, Weiss EI. Streptococcus mutans biofilm changes surface-topography of resin composites. Dent Mater. 2008;24(6):732–6. https://doi.org/10.1016/j.dental.2007.08.003.
37. Fúcio SBP, Carvalho FG, Sobrinho LC, Sinhoreti MAC, Puppin-Rontani RM. The influence of 30-day-old Streptococcus mutans biofilm on the surface of esthetic restorative materials—an in vitro study. J Dent. 2008;36(10):833–9. https://doi.org/10.1016/j.jdent.2008.06.002.
38. Padovani GC, Fúcio SBP, Ambrosano GMB, Sinhoreti MAC, Puppin-Rontani RM. In situ surface biodegradation of restorative materials. Oper Dent. 2014;39(4):349–60. https://doi.org/10.2341/13-089-C.
39. Bourbia M, Ma D, Cvitkovitch DG, Santerre JP, Finer Y. Cariogenic bacteria degrade dental resin composites and adhesives. J Dent Res. 2013;92(11):989–94. https://doi.org/10.1177/0022034513504436.
40. Borges MAP, Matos IC, Mendes LC, Gomes AS, Miranda MS. Degradation of polymeric restorative materials subjected to a high caries challenge. Dent Mater. 2011;27(3):244–52. https://doi.org/10.1016/j.dental.2010.10.009.
41. Nedeljkovic I, de Munck J, Ungureanu A-A, Slomka V, Bartic C, Vananroye A, et al. Biofilm-induced changes to the composite surface. J Dent. 2017;63:36–43. https://doi.org/10.1016/j.jdent.2017.05.015.
42. Fueki K, Ohkubo C, Yatabe M, Arakawa I, Arita M, Ino S, et al. Clinical application of removable partial dentures using thermoplastic resin. Part II: material properties and clinical features of non-metal clasp dentures. J Prosthodont Res. 2014;58(2):71–84.
43. Najeeb S, Zafar MS, Khurshid Z, Siddiqui F. Applications of polyetheretherketone (PEEK) in oral implantology and prosthodontics. J Prosthodont Res. 2016;60(1):12–9. https://doi.org/10.1016/j.jpor.2015.10.001.

44. Zoidis P, Papathanasiou I, Polyzois G. The use of a modified poly-ether-ether-ketone (PEEK) as an alternative framework material for removable dental prostheses. A clinical report. J Prosthodont. 2016;25(7):580–4. https://doi.org/10.1111/jopr.12325.
45. Hahnel S, Scherl C, Rosentritt M. Interim rehabilitation of occlusal vertical dimension using a double-crown-retained removable dental prosthesis with polyetheretherketone framework. J Prosthet Dent. 2018;119(3):315–8. https://doi.org/10.1016/j.prosdent.2017.02.017.
46. Stawarczyk B, Thrun H, Eichberger M, Roos M, Edelhoff D, Schweiger J, Schmidlin PR. Effect of different surface pretreatments and adhesives on the load-bearing capacity of veneered 3-unit PEEK FDPs. J Prosthet Dent. 2015;114(5):666–73. https://doi.org/10.1016/j.prosdent.2015.06.006.
47. Silla M, Eichenberger M, Stawarczyk B. Polyetherketonketon (PEKK) als Restaurationswerkstoff in der modernen Zahnmedizin: eine Literaturübersicht. Quintessenz Zahntech. 2016;42(2):176–90.
48. Kurtz SM, Devine JN. Biomaterials. 2007;28(32):4845–69. https://doi.org/10.1016/j.biomaterials.2007.07.013.
49. Tetelman ED, Babbush CA. A new transitional abutment for immediate aesthetics and function. Implant Dent. 2008;17(1):51–8. https://doi.org/10.1097/ID.0b013e318167648c.
50. Rosentritt M, Ilie N, Lohbauer U, editors. Werkstoffkunde in der Zahnmedizin. Moderne Materialien und Technologien. Stuttgart: Georg Thieme Verlag; 2018.
51. Rosentritt M, Preis V, Behr M, Sereno N, Kolbeck C. Shear bond strength between veneering coposite and PEEK after different surface modifications. Clin Oral Investig. 2015;19(3):739–44. https://doi.org/10.1007/s00784-014-1294-2.
52. Fuhrmann G, Steiner M, Freitag-Wolf S, Kern M. Resin bonding to three types of polyaryletherketones (PAEKs)-durability and influence of surface conditioning. Dent Mater. 2014;30(3):357–63. https://doi.org/10.1016/j.dental.2013.12.008.
53. Stawarczyk B, Eichberger M, Uhrenbacher J, Wimmer T, Edelhoff D, Schmidlin PR. Three-unit reinforced polyetheretherketone composite FDPs: influence of fabrication method on load-bearing capacity and failure types. Dent Mater J. 2015;34(1):7–12. https://doi.org/10.4012/dmj.2013-345.
54. Helkimo E, Carlsson GE, Helkimo M. Bite force and state of dentition. Acta Odontol Scand. 1977;35(6):297–303.
55. Stawarczyk B, Ender A, Trottmann A, Özcan M, Fischer J, Hämmerle CHF. Load-bearing capacity of CAD/CAM milled polymeric three-unit fixed dental prostheses: effect of aging regimens. Clin Oral Investig. 2012;16(6):1669–77. https://doi.org/10.1007/s00784-011-0670-4.
56. Gendreau L, Loewy ZG. Epidemiology and etiology of denture stomatitis. J. Prosthodont. 2011;20(4):251–60.
57. Bilhan H, Sulun T, Erkose G, Kurt H, Erturan Z, Kutay O, Bilgin T. The role of Candida albicans hyphae and Lactobacillus in denture-related stomatitis. Clin Oral Investig. 2009;13(4):363–8. https://doi.org/10.1007/s00784-008-0240-6.
58. Leberer E, Ziegelbauer K, Schmidt A, Harcus D, Dignard D, Ash J, et al. Virulence and hyphal formation of Candida albicans require the Ste20p-like protein kinase CaCla4p. Curr Biol. 1997;7(8):539–46. https://doi.org/10.1016/S0960-9822(06)00252-1.
59. Susewind S, Lang R, Hahnel S. Biofilm formation and Candida albicans morphology on the surface of denture base materials. Mycoses. 2015;58(12):719–27. https://doi.org/10.1111/myc.12420.
60. Ramage G, Coco B, Sherry L, Bagg J, Lappin DF. In vitro Candida albicans biofilm induced proteinase activity and SAP8 expression correlates with in vivo denture stomatitis severity. Mycopathologia. 2012;174(1):11–9. https://doi.org/10.1007/s11046-012-9522-2.
61. Hube B, Albrecht A, Bader O, Beinhauer S, Felk A, Fradin C, et al. Pathogenitätsfaktoren bei Pilzinfektionen. Bundesgesundheitsbl—Gesundheitsforsch—Gesundheitsschutz. 2002;45:159–65.
62. Mendes A, Mores AU, Carvalho AP, Rosa RT, Samaranayake LP, Rosa EAR. Candida albicans biofilms produce more secreted aspartyl protease than the planktonic cells. Biol Pharm Bull. 2007;30(9):1813–5. https://doi.org/10.1248/bpb.30.1813.
63. Gow NAR, van de Veerdonk FL, Brown AJP, Netea MG. Candida albicans morphogenesis and host defence: discriminating invasion from colonization. Nat Rev Microbiol. 2011;10(2):112–22. https://doi.org/10.1038/nrmicro2711.
64. ten Cate JM, Klis FM, Pereira-Cenci T, Crielaard W, de Groot PWJ. Molecular and cellular mechanisms that lead to Candida biofilm formation. J Dent Res. 2009;88(2):105–15. https://doi.org/10.1177/0022034508329273.
65. Douglas LJ. Candida biofilms and their role in infection. Trends Microbiol. 2003;11(1):30–6. https://doi.org/10.1016/S0966-842X(02)00002-1.
66. Pereira-Cenci T, Deng DM, Kraneveld EA, Manders EMM, Del Bel Cury AA, ten Cate JM, Crielaard W. The effect of Streptococcus mutans and Candida glabrata on Candida albicans biofilms formed on different surfaces. Arch Oral Biol. 2008;53(8):755–64. https://doi.org/10.1016/j.archoralbio.2008.02.015.
67. Koch C, Bürgers R, Hahnel S. Candida albicans adherence and proliferation on the surface of denture base materials. Gerodontology. 2013;30(4):309–13. https://doi.org/10.1111/ger.12056.
68. Yoshijima Y, Murakami K, Kayama S, Liu D, Hirota K, Ichikawa T, Miyake Y. Effect of substrate surface hydrophobicity on the adherence of yeast and hyphal

Candida. Mycoses. 2010;53(3):221–6. https://doi.org/10.1111/j.1439-0507.2009.01694.x.
69. Gomes AS, Sampaio-Maia B, Vasconcelos M, Fonesca PA, Figueiral H. In situ evaluation of the microbial adhesion on a hard acrylic resin and a soft liner used in removable prostheses. Int J Prosthodont. 2015;28(1):65–71. https://doi.org/10.11607/ijp.4080.
70. Tari BF, Nalbant D, Al Dogruman F, Kustimur S. Surface roughness and adherence of Candida albicans on soft lining materials as influenced by accelerated aging. J Contemp Dent Pract. 2007;8(5):18–25.
71. Freitas-Fernandes FS, Cavalcanti YW, Ricomini Filho AP, Silva WJ, Del Bel Cury AA, Bertolini MM. Effect of daily use of an enzymatic denture cleanser on Candida albicans biofilms formed on polyamide and poly(methyl methacrylate) resins: an in vitro study. J Prosthet Dent. 2014;112(6):1349–55. https://doi.org/10.1016/j.prosdent.2014.07.004.
72. O'Donnell LE, Smith K, Williams C, Nile CJ, Lappin DF, Bradshaw D, et al. Dentures are a reservoir for respiratory pathogens. J Prosthodont. 2016;25(2):99–104. https://doi.org/10.1111/jopr.12342.
73. Urushibara Y, Ohshima T, Sato M, Hayashi Y, Hayakawa T, Maeda N, Ohkubo C. An analysis of the biofilms adhered to framework alloys using in vitro denture plaque models. Dent Mater J. 2014;33(3):402–14.
74. El-Solh AA. Association between pneumonia and oral care in nursing home residents. Lung. 2011;189(3):173–80. https://doi.org/10.1007/s00408-011-9297-0.
75. Iinuma T, Arai Y, Abe Y, Takayama M, Fukumoto M, Fukui Y, et al. Denture wearing during sleep doubles the risk of pneumonia in the very elderly. J Dent Res. 2015;94(3 Suppl):28S–36S. https://doi.org/10.1177/0022034514552493.
76. Niederman MS. Nosocomial pneumonia in the elderly patient. Chronic care facility and hospital considerations. Clin Chest Med. 1993;14(3):479–90.
77. Sjögren P, Nilsson E, Forsell M, Johansson O, Hoogstraate J. A systematic review of the preventive effect of oral hygiene on pneumonia and respiratory tract infection in elderly people in hospitals and nursing homes: effect estimates and methodological quality of randomized controlled trials. J Am Geriatr Soc. 2008;56(11):2124–30. https://doi.org/10.1111/j.1532-5415.2008.01926.x.
78. Mojon P. Oral health and respiratory infection. J Can Dent Assoc. 2002;68(6):340–5.
79. Scannapieco FA, Papandonatos GD, Dunford RG. Associations between oral conditions and respiratory disease in a national sample survey population. Ann Periodontol. 1998;3(1):251–6. https://doi.org/10.1902/annals.1998.3.1.251.
80. Tada A, Miura H. Prevention of aspiration pneumonia (AP) with oral care. Arch Gerontol Geriatr. 2012;55(1):16–21. https://doi.org/10.1016/j.archger.2011.06.029.
81. Astasov-Frauenhoffer M, Glauser S, Fischer J, Schmidli F, Waltimo T, Rohr N. Biofilm formation on restorative materials and resin composite cements. Dent. Mater. 2018; https://doi.org/10.1016/j.dental.2018.08.300.
82. Aykent F, Yondem I, Ozyesil AG, Gunal SK, Avunduk MC, Ozkan S. Effect of different finishing techniques for restorative materials on surface roughness and bacterial adhesion. J Prosthet Dent. 2010;103(4):221–7. https://doi.org/10.1016/S0022-3913(10)60034-0.
83. Eick S, Glockmann E, Brandl B, Pfister W. Adherence of Streptococcus mutans to various restorative materials in a continuous flow system. J Oral Rehabil. 2004;31(3):278–85. https://doi.org/10.1046/j.0305-182X.2003.01233.x.
84. Tanner J, Robinson C, Söderling E, Vallittu P. Early plaque formation on fibre-reinforced composites in vivo. Clin Oral Investig. 2005;9(3):154–60.
85. Hauser-Gersprach I, Kulik EM, Weiger R, Decker E-M, von Ohle C, Meyer J. Adhesion of Streptococcus sanguinis to dental implant and restorative materials in vitro. Dent Mater J. 2007;26(3):361–6. https://doi.org/10.4012/dmj.26.361.

7 Effect of Oral Biofilms on Dental Materials: Biocorrosion and Biodeterioration

Ivana Nedeljkovic

Abstract

By restoring decayed, traumatized, or missing tooth tissues, dentists introduce a new substrate into the oral cavity, to which dental biofilm (plaque) can adhere and accumulate. Even though microbial adhesion and biofilm development and maturation on a dental material surface follow some general patterns, these processes also depend on the properties of the material itself. There are specific interactions between dental materials and the overlying biofilms. On the one hand, materials can directly affect biofilms by releasing bioactive compounds, which gives an opportunity for the biofilm control and the prevention of secondary caries and other oral infectious diseases. On the other hand, dental plaque has a potential to modify the restorative material's surface properties, such as surface roughness and topography, which might boost bacterial accumulation and eventually compromise restoration's longevity. Resin-based composites, which are the most commonly used restorative materials nowadays, seem to be particularly prone to biofilm-induced degradation, since a well-known cariogenic species, *Streptococcus mutans*, can produce enzymes with esterase activity, capable of breaking down the polymer matrix of composites. However, the regulatory mechanisms behind the production and activity of such enzymes within a large community of different species in dental plaque remain obscure.

7.1 Introduction

7.1.1 Biofilms and Dental Materials: General Terms

As mentioned already many times throughout this book, biofilms are surface-associated aggregates or communities of microbial cells, which are embedded in a self-produced extracellular matrix (ECM) or extracellular polymeric substance (EPS) [1]. The fact that they are developing at a surface or an interface distinguishes biofilm microbial cells substantially from their planktonic or free-living counterparts. First, in order to initially attach to the surface, microbial cells need to express phenotypes which would allow them to do so. Initial attachment allows cells to stay in close proximity and to start interacting with the cells from the same, but also from different bacterial species, thereby developing a complex multicellular community, in which they express a number of new, so-called emergent properties [2]. One of these properties is the production of ECM, which is essential for the structure and functioning of a biofilm.

I. Nedeljkovic (✉)
Academic Center for Dentistry Amsterdam (ACTA), Amsterdam, The Netherlands

A. C. Ionescu, S. Hahnel (eds.), *Oral Biofilms and Modern Dental Materials*,
https://doi.org/10.1007/978-3-030-67388-8_7

Biofilms are one of the most ubiquitous modes of life on Earth. They colonize soil, natural aquatic systems, and all higher organisms including humans, but also industrial and potable water systems, medical devices, ship hulls, etc. It seems that solid-liquid interfaces between a solid surface and an aqueous medium are particularly suitable for biofilm development, due to their constant need for hydration [1]. One of these kinds of habitats is definitely an oral cavity. Oral cavity is a host of more than 700 microbial species, which grow either as planktonic cells or in the form of biofilms, widely known as dental plaque, developing on oral soft and hard tissues [3, 4]. In addition, dental plaque may develop on the surface of a wide variety of dental materials introduced into an oral cavity as a part of oral rehabilitation.

Even though biofilms have a number of important functions in nature, such as biogeochemical cycling and symbiotic relationship with a number of plant and animal species, including humans, their presence often has negative effects. Accumulation of different micro- (and macro-) organisms on the wetted surfaces or solid-liquid interfaces, commonly known as biofouling, might have a deleterious effect on the underlying surfaces themselves, as well as on the whole systems (artificial or living) these surfaces are part of [5]. Oral cavity is no exception here. Even though the oral microbiome, as a part of the whole human microbiome, plays a critical role in many metabolic, physiological, and immunological processes, such as maturation and differentiation of host mucosa and immune system, food digestion and nutrition, and protection from pathogenic microorganisms, it can, under certain circumstances, cause some of the most prevalent dental diseases such as caries, gingivitis, and periodontitis [4]. In addition, plaque accumulation on dental restorative materials can have a negative impact not only on the surrounding tissues, but also on the underlying materials, which might seriously affect their clinical performance.

As soon as they are introduced in the oral cavity, dental materials start interacting with oral bacteria. These interactions are of paramount importance in the process of the bacterial adhesion and biofilm formation and accumulation on the materials, and they remain important throughout the whole service of the material in mouth. Dental restorative materials include a broad spectrum of materials, such as metals and alloys, amalgams, ceramics, polymers, and composites, all of which interact with oral biofilms in a distinct manner. The interaction is per definition a mutual or reciprocal action or influence (*Merriam-Webster's dictionary*). Since the effect of various bioactive and antibacterial materials on the biofilms has already been discussed in the previous few chapters, the main focus of this chapter will be the manners in which oral biofilms can affect the underlying restorative materials and the performance of dental restorations. Also, the focus will mainly be laid on the most commonly used dental materials for direct dental restorations, namely dental composites and dental amalgams. Even though dental amalgams have been the standard restorative for more than a century, their use has been gradually abandoned in many developed countries due to their poor esthetics, as well as health and environmental concerns [6, 7]. Dental composites, on the other hand, have become a gold standard for dental restorations, due to their ability to adhesively bond to a tooth, thereby supporting the preservation of healthy tooth tissues, their versatility, esthetics, easy handling, etc. Nevertheless, it seems that conventional dental composites have a shorter longevity and a higher replacement rate than amalgams, which has mainly been attributed to their higher susceptibility to secondary or recurrent caries (SC) [8–10]. In addition, composites seem to be less resistant to the degradation in a quite challenging environment such as the oral cavity, especially by biological factors including oral bacteria [11]. This has brought restorative material-biofilm interactions into the spotlight, since it could give a better insight into the secondary caries process and failures of different restorations and help improve new generations of restorative materials.

The term "biodeterioration" is suitably used when talking about the impairment of function and/or esthetic properties of synthetic polymer materials by microorganisms [5]. This is mainly done through the decomposition of the polymer chains by the microbial activity, or so-called biodegradation. In the remainder of this chapter, these terms will be used to discuss biofilm-induced alterations of dental composites and adhesives, since their organic component, a resinous matrix, which actually undergoes degradation, is a polymer in chemical terms. On the other hand, dental amalgams are composed of a mixture of metal alloys, and microbial deposition on amalgam restorations might induce a corrosion process, known as "biocorrosion."

7.1.2 Relevant Aspects of Biofilms

Biofilm formation on hard oral surfaces, including dental materials, follows a general pattern and consists of the following steps: acquired pellicle formation, initial bacterial cell attachment or so-called primary/early colonization, secondary colonization or co-aggregation, and biofilm maturation, which could be followed by cell detachment and dispersion [12, 13]. This process is, however, affected by many factors, including a number of environmental and host factors (temperature, pH, oxygen levels, nutrient availability, shear stresses, antimicrobial peptides, etc.) and bacterial cell factors (hydrophobicity, presence of fimbriae and flagella, production of EPS), but also various properties of the substrate (surface roughness and topography, stiffness, charge, hydrophobicity, chemical composition) [1, 14]. It is thus no wonder that both quantitative and qualitative differences in biofilms growing on different dental materials have been reported. It has been shown that conventional composites accumulate more biofilms on their surface compared to amalgams and glass ionomer cements [15]. In addition, it seems that plaque growing on composites contains a higher proportion of cariogenic species, such as mutans streptococci and lactobacilli [16, 17]. This could be explained by the lack of antibacterial properties of composites compared to other two restoratives, or by the lack of buffering or pH-neutralizing abilities [18]. Nevertheless, it can definitely make composites more exposed and more susceptible to biodegradation, especially by cariogenic species, which as a matter of fact seem to have a higher biodegradation potential.

Irrespective of the substrate, the basic structure of mature biofilms includes densely packed microbial cells (from 10^8 to 10^{11} cells per gram wet weight), self-produced extracellular polymeric substance (EPS), and interstitial pores and channels which facilitate transport of water and metabolites [2]. EPS comprises the largest part of the biofilm mass (75–95%) and is of the greatest importance in the interactions between the biofilm and the substrate [13]. EPS mediates the biofilm growth at the surface of the substrate and it imparts various important properties to the biofilms, such as resource capture by sorption, enzyme retention and digestive capacities, intercellular interactions (competition and cooperation) and metabolism, and resistance to desiccation and antimicrobials.

EPS is actually capable of retaining and stabilizing extracellular enzymes secreted by bacterial cells, which allows it to function as a sort of an external digestive system. In this way, the concentration and thereby the efficacy of bacterial enzymes are substantially higher than in case of planktonic cells, where enzymes easily diffuse and get diluted after the secretion. This enzymatic system is important for the digestion of the nutrients taken up from the environment, but it also allows biofilm to attack the substrate it is attached to, as it will be discussed later in this chapter. In addition, processes of sorption and accumulation of various compounds from the environment and the compounds released from the substrate play an important role in the modulation of bioactivity and toxicity of dental materials. These two important functions of the EPS are schematically presented in Fig. 7.1 [2].

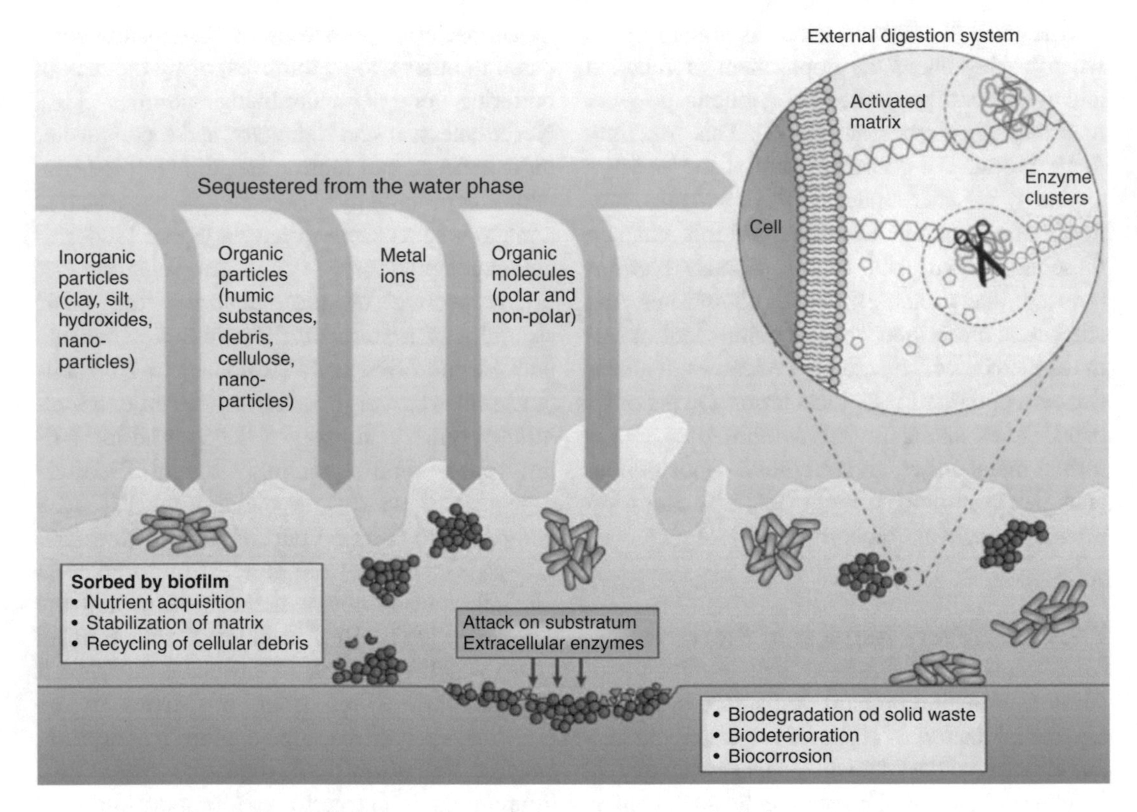

Fig. 7.1 Capturing and retention of external resources as well as extracellular bacterial enzymes by EPS in biofilms [2]. (Permission obtained from Springer Nature, license number 4710220788887)

7.2 Biofilms and Amalgam Restorations

Dental amalgams have been for a long time considered a gold standard among restorative materials. Nevertheless, during the last two decades the use of amalgams has been on a steady decline, and in many developed countries it is nowadays merely used, or even banned, due to health and environmental concerns [19]. Minamata Convention on Mercury (2013) is an international treaty, which proposed a number of measures to decrease anthropogenic emission and release of mercury, including the phasedown of dental amalgams, and their replacement with mercury-free alternatives. However, due to a relatively simple and insensitive placement technique, high longevity, and unparalleled cost-effectiveness of amalgams, they are still widely used, especially in low- and middle-income countries [20].

Amalgams are alloys of mercury and other metals, such as silver, tin, copper, and metallic elements added to improve their physical and mechanical properties (ADA, 2011). Dental amalgams are the only metallic materials for direct tooth restorations, and their interactions with the oral environment differ substantially from the interactions of dental composites or glass ionomer cements, which both consist of inorganic as well as organic components. Amalgams have arguably the highest longevity among direct restoratives, and it seems that they perform particularly better compared to composites in patients with high caries risk [21]. A closer look into the specific interactions between amalgams and oral biofilms could perhaps offer an explanation for their higher resilience in oral cavity.

Accumulation of oral biofilms on the surface of dental amalgams has a potential to cause a bacterium-induced corrosion or so-called biocor-

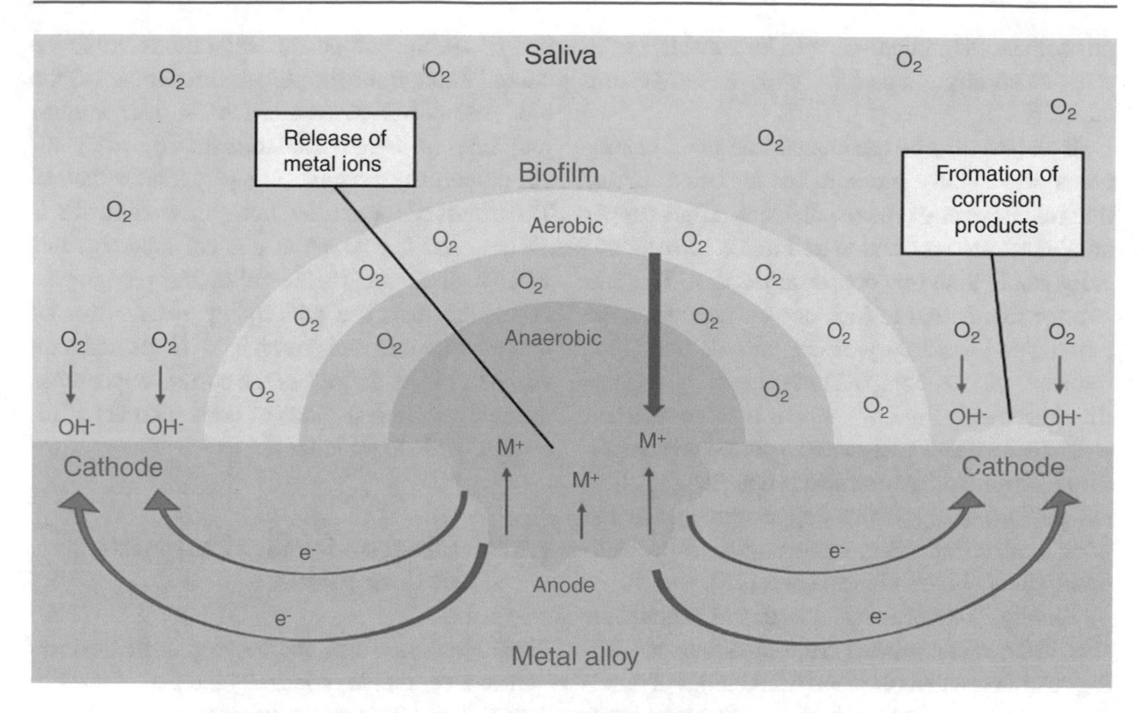

Fig. 7.2 The schematic presentation of the mechanisms of biofilm-induced corrosion

rosion. With regard to its mechanism, biocorrosion belongs to concentration cell type of corrosion, which is an electrochemical corrosion that occurs when there is a difference in the electrolyte composition within one system. For instance, when the surface of an alloy is covered by biofilm or another kind of debris, which can produce an electrolyte different from the one at the rest of the surface (saliva), corrosion might take place. Accumulation of bacterial metabolic products, including organic acids, causes the drop in local pH, which together with the depletion of oxygen leads to the formation of a corrosion cell. This further causes the release of metal ions and the formation of corrosion products [22] (Fig. 7.2). This process is especially accelerated in surface defects, such as pits and cracks, since these areas are oxygen deprived. It is therefore important to polish amalgam restorations, in order to obtain a smooth and homogenous surface less prone to plaque accumulation and corrosion.

Biocorrosion has a two-sided effect on dental amalgam restorations. On the one hand, it can lead to the release of free metal ions from amalgam, especially zinc, copper, and tin, and at much lower rate silver and mercury [23]. The release of metal ions due to corrosion process is important from the aspect of biocompatibility and toxicity of amalgams, and it has therefore been a research focus for a long time. It appears that the presence of biofilms at the surface of amalgam restorations plays another role here since they can capture and accumulate the released ions by sorption process, which actually retards their release into the oral environment [24]. On the other hand, corrosion leads to the formation of solid nonmetallic compounds, such as oxides, hydroxides, and chlorides of tin, copper, and zinc. These products mostly stay bound to amalgam structure or form a layer on top of it. Even though the formation of corrosion products might affect the mechanical properties of amalgams and lead to increased abrasion and fragmentation, there is also a positive aspect of it. Namely, the formation of corrosion products at amalgam-tooth interface, where a so-called crevice corrosion often occurs, can seal the interfacial gap and prevent microleakage and its consequences, such as postoperative sensitivity and secondary caries [25]. Owing to this phenomenon, amalgams could be considered the only restorative materials with a self-sealing capacity.

It should also be mentioned that not all amalgams are equally prone to biocorrosion, or for that matter to any type of corrosion. High-copper amalgams are reported to be more electrochemically stable than low-copper amalgams. The reason for this is that in high-copper amalgams the amount of tin-mercury phase (gamma two, γ_2) is reduced, or it is completely replaced by copper-tin phase (eta prime, η'), which is more resistant to corrosion. This resistance to corrosion contributes to improved mechanical properties and clinical performance of high-copper amalgams, but what is also important, it does not affect considerably their self-sealing abilities [25].

Finally, biocorrosion of dental amalgams should be distinguished from amalgam tarnishing, which is a discoloration (darkening) of amalgam restoration surface due to the formation of a thin, adherent, and insoluble film at its surface, consisting mainly of silver and copper sulfides. Although they negatively affect the esthetics of amalgam restorations (loss of luster), they do not seem to affect mechanical and functional properties of amalgams, and are even considered to act protectively against the corrosion.

7.3 Biodeterioration of Dental Composites

As mentioned above, dental composites seem to accumulate more plaque compared to dental amalgams and glass ionomer cements. This makes them in a way more exposed to the adverse effects of biofilms or biodeterioration. Nevertheless, dental composites are the most commonly used restorative materials and it is of utmost importance to have a better understanding of the damage they might suffer due to plaque accumulation and the biodegradation processes taking place in oral cavity.

Biodeterioration of composites is considered to be able to seriously compromise the function and longevity of composite restorations, and it has been therefore extensively investigated during the last decade. It appears that various properties of dental composite restorations could be altered through the biofilm accumulation and the biodeterioration process, such as their surface properties (roughness and topography), mechanical properties, marginal integrity, and esthetics. The extent of these alterations, their clinical relevance, and the potential clinical repercussions will be discussed in the following paragraphs. Following that, the underlying mechanisms of biodeterioration and composite biodegradation will be tackled, as well as the current approaches to improve the resistance of contemporary composites to biodegradation.

7.3.1 Effect on Surface Properties of Composites

Since biofilms attach and develop at the restoration surface, that is expectedly the part of a restoration first affected by microbial degradation. Influence of biodegradation on the surface properties of composites, such as surface roughness, topography, and surface hardness, has therefore been most extensively investigated in literature to date. Several studies demonstrated that cariogenic species *S. mutans* can degrade the surface of dental composites, and thereby increase the surface roughness and change the surface topography [26, 27]. This has gained a lot of attention since the increase in surface roughness can boost further bacterial accumulation. Nevertheless, it seems that this effect depends on the type of biofilm used in in vitro studies. Gregson et al. (2012) have shown that cariogenic species *S. mutans* can noticeably change surface topography of composites and increase its roughness, while a non-cariogenic species *S. sanguinis* does not exhibit the same biodegradation potential [28]. In another in vitro study a significant increase in surface roughness of two composites after 6-week incubation with *S. mutans* single-species biofilms was found, while there was no significant increase in roughness after the exposure to a multispecies model, consisting of *S. mutans*, *S. sanguinis*, *A. naeslundii*, and *F. nucleatum* [29] (Fig. 7.3). Even though it was shown that few other oral species, such as *S. gordonii* and *A. naeslundii*, also have the ability to degrade the composite surface

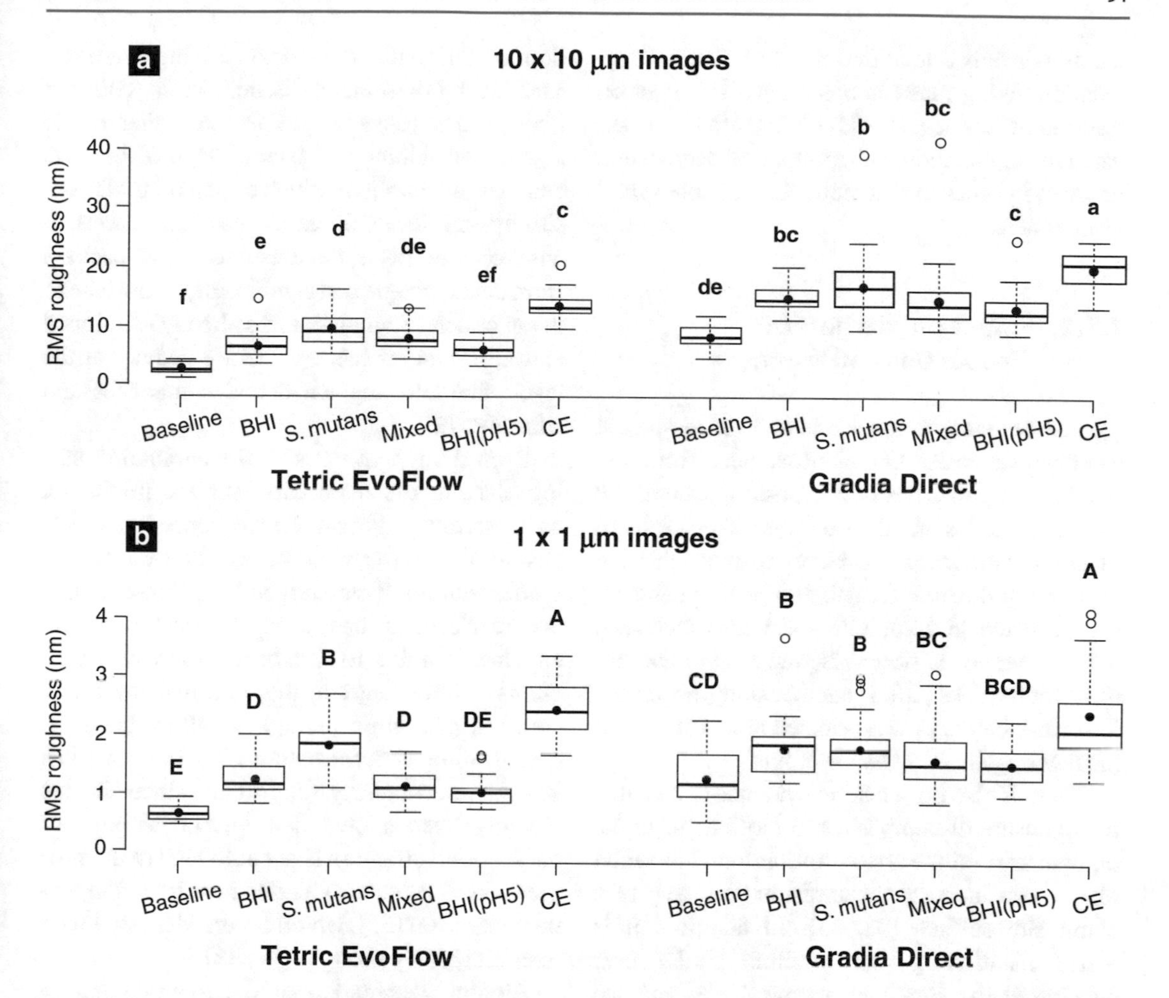

Fig. 7.3 Root mean square (RMS) roughness values of the samples of two composites measured by atomic force microscopy (AFM) at lower (**a**) and higher (**b**) resolution, at baseline (**Baseline**), and after 6-week incubation under the following conditions: in sterile bacterial growth medium (**BHI**), with *S. mutans* biofilm (***S. mutans***), with mixed four-species biofilm (**Mixed**), in sterile bacterial growth medium with pH adjusted to 5 (**BHI(pH 5)**), and in cholesterol esterase solution in PBS (**CE**) [29]. (Permission obtained from Elsevier, license number 4710221190144)

and increase its roughness [28, 30], most of the studies are focused exclusively on the "old villain" *S. mutans*. However, in the light of the latest research, which suggests that biodegradation potential of *S. mutans* diminishes when co-cultured with other species, clinical relevance of the in vitro studies on composite biodegradation using single-species models with *S. mutans* should be questioned, especially considering the fact that dental plaque is a community of more than 700 species.

Another critical question here is whether the increase in surface roughness observed in the abovementioned in vitro studies is clinically relevant and whether it can actually lead to an increase in bacterial accumulation. Teughels et al. tried to determine a critical value of surface roughness above which a significant increase in bacterial accumulation can be observed [31]. The obtained average roughness (R_a) value of 200 nm (which corresponds to root mean square (RMS) value of 220 nm) is way much higher than the roughness measured on biofilm-exposed composite surfaces in in vitro studies, which approximately ranged from 10 to 50 nm [26, 29]. It therefore seems that the ability of certain oral bacterial species, such as *S. mutans*, to degrade the surface of dental composites has no potential to seriously compromise a clinical performance of composite restorations. It should however be kept in mind that bacterial degradation is only one of the modes of material degradation taking place in oral cavity. Its effect on restoration surface should be therefore investigated in combination with mechanical degradation or wear, in order to determine the possible synergistic effects.

7.3.2 Effect on Mechanical Properties and Wear

Apart from surface roughness and topography, it has been suggested that biofilms can affect other mechanical properties of composites, such as the surface hardness and the wear rate. The results of several in vitro studies, however, disputed this. No decrease in flexural strength and surface hardness was detected in composite specimens incubated with *S. mutans*, *S. sanguinis*, and *S. gordonii* biofilms for 6 weeks [28]. In another study, no change in surface hardness was detected after 1 month of incubation with *S. mutans* biofilm [26].

There is also no scientific evidence to date that the exposure of composites to biofilms or to the organic acids at the concentration found in dental plaque can increase the abrasion and wear of a composite surface [32, 33]. In addition, it is worth mentioning that biofilms hardly ever develop at the sites which typically experience wear, such as occlusal surfaces.

Based on the present literature it could be concluded that bacterial degradation of dental composites takes place at and is limited to the outer material surface, without affecting materials' inner (bulk) structure and thereby their mechanical qualities.

7.3.3 Effect on Tooth-Composite Interface

Integrity of the tooth-restoration interface is crucial to achieving high longevity and optimal clinical performance of composite restorations. It seems, however, that during restoration service in oral cavity this interface may considerably deteriorate, which may eventually lead to the restoration failure [34]. Deterioration is a consequence of mechanical as well as biochemical degradation of different components of the interface, such as tooth mineral tissue, dentin collagen fibers, and adhesive layer. This can further lead to a so-called micro- and nanoleakage of bacteria and their metabolites, which can cause tooth sensitivity and development of secondary caries. It has recently been demonstrated that bacteria from dental plaque can significantly contribute to the interfacial degradation. As already mentioned above, *S. mutans* has an esterase activity at the levels that can degrade dental composites and adhesives [27].

Even though composite biodegradation taking place at the restoration surface might not have serious clinical consequences, as discussed above, the same cannot be said for the biodegradation happening at the tooth-composite interface. It has been demonstrated that enzymes similar to the ones produced by *S. mutans* could degrade the adhesive layer and create a gap large enough to allow bacterial colonization and formation of a biofilm [35]. This is particularly important since it has recently been shown that interfacial gaps of only around 30 μm in size could lead to the progression of secondary caries next to a composite restoration [36], which is considerably lower than previously thought [37, 38].

Biofilm degradation of the tooth-composite interface is also reflected on the bond strength between dentin and composite. Li et al. demonstrated a reduction in bond strength after specimen exposure to multispecies biofilms, especially in the presence of sucrose in the growth medium [39]. The observed reduction in bond strength has not been attributed only to the dentin demineralization, but also to the hydrolysis of the adhesive resin by either bacterium-produced acids or bacterium-produced enzymes.

7.3.4 Effect on Esthetic Properties of Composites

Excellent esthetic properties are one of the greatest assets of dental composites, and one of the main reasons for their high popularity among patients and dentists. Nevertheless, during their service in the mouth, the appearance of composite restorations can significantly deteriorate, and the discoloration they undergo can be per se a reason for the restoration replacement, especially in the esthetic zone. The percentage of composite restorations replaced due to bulk and marginal discoloration has been reported to range from 3% to as high as 22%, and was often reported to be the second or third most common reason for the replacement [40–42]. Furthermore, it appears that the staining of composite fillings is associated with patient's poor oral hygiene, and accumulation of oral biofilms is often stated as an important intrinsic factor affecting color stability of composites [43, 44].

Nevertheless, the literature about the direct influence of biofilms on the deterioration of esthetic properties of composites is quite scarce. A relatively recent study investigated the effect of *S. mutans* biofilm on the color and translucency of experimental composites with and without bioactive glass fillers [45]. The results showed no difference in the change of optical properties after the exposure of the control composite to *S. mutans* culture and to the growth medium alone, which implies the absence of any direct effect of bacteria whatsoever. *S. mutans* biofilms were used in this study because of the high production of acids, which were long considered contributing factors to the color change of composites. There is, however, no sound scientific evidence for that, and more research with multispecies biofilm models is needed to get a better insight into direct effects of oral bacteria on optical properties of composites.

On the other hand, biodegradation of composites and adhesives could affect esthetics of composite restorations indirectly. An increased roughness of composite surface also means a larger surface for the adsorption of pigments from foods and beverages. It has been shown that different polishing techniques and initial roughness of composites can influence their color stability, but it seems that this depends to a great extent on the type of composite material [46]. Also, biodegradation of the adhesive bond can lead to the leakage and accumulation of pigments at the tooth-composite interface and cause marginal discoloration, which, as already mentioned, can be a reason for the restoration replacement.

7.3.5 Mechanisms of Bacterial Degradation

Composite biodegradation is based on the hydrolysis of the chemical bonds present in resin polymer matrix, such as ester, urethane, and amide bonds. Hydrolytic reaction can be catalyzed or facilitated by acids, bases, and also different enzymes, when we talk about enzymatic hydrolysis and enzymatic degradation. Bacteria from dental biofilms are known to be able to efficiently produce organic acids under cariogenic challenge, especially so-called cariogenic bacterial species, such as mutans streptococci and lactobacilli, which are present at higher proportions in cariogenic biofilms. Therefore, it has long been considered that the main mechanism of microbial degradation of composites in oral cavity is an acid-catalyzed hydrolysis [47]. During the recent years, however, it has been demonstrated that *S. mutans* species can produce enzymes from the class of esterases, similar to cholesterol esterase (CE) and pseudocholinesterase (PCE) found in saliva, which are able to degrade methacrylate monomers within composite matrix, such as TEGDMA and BisGMA. Moreover, the produced esterase remains stable and active even at low pH level of 5.5, which is found in cariogenic plaque [48]. The mechanism of microbial degradation of resin composites and adhesives can be thus regarded as a combination of acid- and enzyme-catalyzed hydrolytic degradation. Nevertheless, a recent study, which investigated the effect of biofilms on the surface of resin composites, suggested that bacterial enzymes, rather than acids, play a role in microbial degradation of composites, since no effect of bacterial growth medium with low pH (pH = 5) on the tested composites was observed [29] (Fig. 7.3).

Even though bacterium-produced acids have little contribution to the degradation of composite surface, their role in the interfacial degradation seems to be quite prominent. Apart from the demineralization of tooth mineral tissues, which

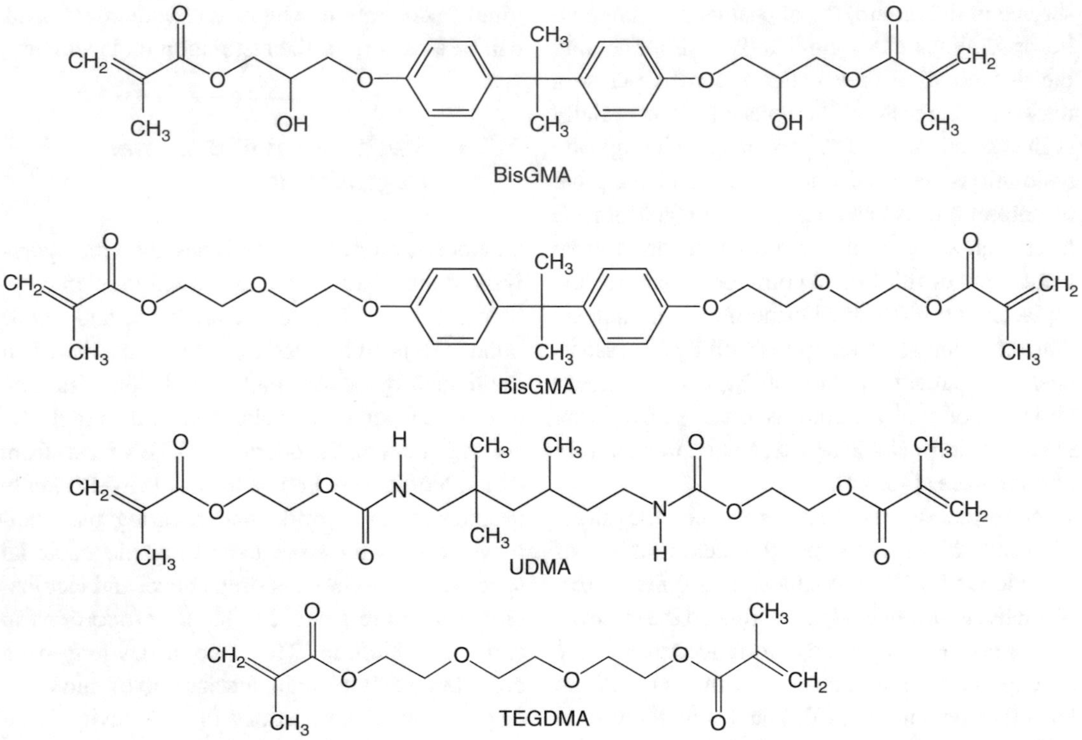

BisGMA

BisGMA

UDMA

TEGDMA

Fig. 7.4 Structural formulas of monomers most commonly found in dental composites

is an important aspect of interfacial degradation, bacterium-produced acids could also be responsible for the activation of certain proteolytic enzymes present in saliva and in dentin. These enzymes are known as matrix metalloproteinases (MMPs) and cysteine cathepsins, and they are considered to play a part in the interfacial breakdown by degrading collagen fibrils in hybrid layer [49].

7.3.6 Susceptibility/Resistance to Biodegradation

Not all composites are equally prone to hydrolytic degradation, and the susceptibility to degradation largely depends on the material composition. In the first instance it is determined by the silanated filler fraction, as highly filled composites show a higher resistance to the biodegradation than composites with a lower filler content. This is no surprise considering the fact that the resin matrix is a vulnerable component of composites when it comes to chemical degradation, and in highly filled composites a smaller matrix surface is exposed to the activity of enzymes [50].

In addition, susceptibility to degradation is determined by the resin matrix chemistry, as certain types of resin monomers are more prone to hydrolysis than others. Ester bonds, which are present in most of the currently used monomers, are particularly susceptible to degradation. However, the presence of other chemical groups on monomer molecules and their interactions can affect their stability considerably. Among most commonly used resin monomers, which are presented in Fig. 7.4, triethylene glycol dimethacrylate (TEGDMA) seems to be the most susceptible to degradation [51]. A possible reason for this is the presence of ethylene glycol segments, which attract water molecules and increase the water uptake, leading to a higher chance for hydrolysis [11]. Aromatic cross-linking monomer bisphenol A-glycidyl methacrylate (BisGMA) and its ethoxylated version (BisEMA) are more stable than

TEGDMA, due to the presence of hydrophobic aromatic rings in their backbone, which partly protect polar groups from water and hydrolysis. Nevertheless, their susceptibility to hydrolytic degradation is still quite high [52]. On the other hand, monomers containing urethane groups, such as urethane dimethacrylate (UDMA), but also urethane-modified BisGMA, show considerably lower susceptibility to degradation compared to other monomers present in contemporary composite materials [51, 52]. Urethane groups can form hydrogen-bonded structures which can restrict the access of enzymes to the cleavage sites, thereby delaying enzymatic reaction and protecting ester bonds in their vicinity from the hydrolysis. In addition, the elimination of hydroxyl groups by the formation of urethane links in urethane-modified BisGMA leads to an increased hydrophobicity of the monomer and a higher resistance to hydrolytic attack.

In the last years, much research has been devoted to designing new monomers with different chemistries, which would have a higher resistance to biodegradation in oral cavity. Several studies reported quite promising results with experimental monomers for composites, as well as for adhesive resins. Gonzalez-Bonet et al. synthesized and tested an ether-based monomer triethylene glycol divinylbenzyl ether (TEG-DVBE), which showed no signs of degradation in PBS, cholesterol esterase (CE), and pseudocholine esterase (PCE) solutions, compared with BisGMA and TEGDMA, which degraded at different levels [53]. Another group tested a quaternary methacrylamide-based ammonium fluoride and demonstrated a high resistance of this antibacterial monomer to hydrolysis in acidic environment [54].

7.4 Conclusions

Interactions between dental restorative materials and oral biofilms might be an important determinant of their clinical performance. Certain qualities of restorative materials, such as antibacterial and pH-neutralizing effect and lower plaque accumulation, but also higher resistance to bacterial degradation and biodeterioration in the oral cavity, can contribute to an improved longevity of dental restorations. This can explain superior longevity and resistance to secondary caries of dental amalgams, which do not seem to be adversely affected by oral biofilms. They can even benefit from the biocorrosion, since the solid by-products of corrosion can seal the gap at the tooth-restoration interface, thereby preventing microleakage and development of secondary caries.

On the other hand, dental composites seem to be more vulnerable to biodeterioration, which might affect various composite properties. The effect of biofilms on surface roughness and mechanical properties, such as surface hardness and wear, seems not to pose a clinical problem. However, bacterial degradation can contribute to the disintegration of the tooth-composite interface in multiple ways, including the breakdown of the adhesive layer, tooth mineral tissues, and dentin collagen. Interfacial degradation, in its turn, can lead to marginal discoloration and deterioration of composite's appearance. Nevertheless, the direct effect of oral biofilms, as well as the combined effect of biofilms and exogenous discoloring factors on esthetic properties of composite restorations, has been scarcely investigated and is still not clear.

Resin chemistry plays a crucial role in the resistance of composites to bacterial degradation. The research on designing new, more biochemically stable formulations of resin monomers is gaining increasing attention, and encouraging results have already been reported. Improved biostability of dental composites would help to improve their clinical performance and prolong their service in mouth.

References

1. Donlan RM. Biofilms: microbial life on surfaces. Emerg Infect Dis. 2002;8(9):881–90.
2. Flemming HC, et al. Biofilms: an emergent form of bacterial life. Nat Rev Microbiol. 2016;14(9):563–75.
3. Samaranayake L, Matsubara VH. Normal oral flora and the oral ecosystem. Dent Clin N Am. 2017;61(2):199–215.
4. Kilian M, et al. The oral microbiome—an update for oral healthcare professionals. Br Dent J. 2016;221(10):657–66.

5. Flemming HC. Biodeterioration of synthetic materials—a brief review dedicated to Professor Dr. Wolfgang Sand on the occasion of his 60th birthday. Mater Corros—Werkstoffe Und Korrosion. 2010;61(12):986–92.
6. Forss H, Widstrom E. From amalgam to composite: selection of restorative materials and restoration longevity in Finland. Acta Odontol Scand. 2001;59(2):57–62.
7. Sunnegardh-Gronberg K, et al. Selection of dental materials and longevity of replaced restorations in Public Dental Health clinics in northern Sweden. J Dent. 2009;37(9):673–8.
8. Opdam NJ, et al. A retrospective clinical study on longevity of posterior composite and amalgam restorations. Dent Mater. 2007;23(1):2–8.
9. Bernardo M, et al. Survival and reasons for failure of amalgam versus composite posterior restorations placed in a randomized clinical trial. J Am Dent Assoc. 2007;138(6):775–83.
10. Moraschini V, et al. Amalgam and resin composite longevity of posterior restorations: a systematic review and meta-analysis. J Dent. 2015;43(9):1043–50.
11. Delaviz Y, Finer Y, Santerre JP. Biodegradation of resin composites and adhesives by oral bacteria and saliva: a rationale for new material designs that consider the clinical environment and treatment challenges. Dent Mater. 2014;30(1):16–32.
12. Marsh PD. Dental plaque as a biofilm and a microbial community—implications for health and disease. BMC Oral Health. 2006;6(Suppl 1):S14.
13. Huang R, Li M, Gregory RL. Bacterial interactions in dental biofilm. Virulence. 2011;2(5):435–44.
14. Song F, Koo H, Ren D. Effects of material properties on bacterial adhesion and biofilm formation. J Dent Res. 2015;94(8):1027–34.
15. Zhang N, et al. Do dental resin composites accumulate more oral biofilms and plaque than amalgam and glass ionomer materials? Materials (Basel). 2016;9(11):888.
16. Svanberg M, Mjor IA, Orstavik D. Mutans streptococci in plaque from margins of amalgam, composite, and glass-ionomer restorations. J Dent Res. 1990;69(3):861–4.
17. Thomas RZ, et al. Bacterial composition and red fluorescence of plaque in relation to primary and secondary caries next to composite: an in situ study. Oral Microbiol Immunol. 2008;23(1):7–13.
18. Nedeljkovic I, et al. Lack of buffering by composites promotes shift to more cariogenic bacteria. J Dent Res. 2016;95(8):875–81.
19. Kopperud SE, et al. The post-amalgam era: Norwegian dentists' experiences with composite resins and repair of defective amalgam restorations. Int J Environ Res Public Health. 2016;13(4):441.
20. Fisher J, et al. The Minamata convention and the phase down of dental amalgam. Bull World Health Organ. 2018;96(6):436–8.
21. Opdam NJ, et al. 12-year survival of composite vs. amalgam restorations. J Dent Res. 2010;89(10):1063–7.
22. Anusavice KJ, Phillips RW. Phillips' science of dental materials. 11th ed. St. Louis, MO: Saunders; 2003. p. xxv, 805 p.
23. Marek M. Interactions between dental amalgams and the oral environment. Adv Dent Res. 1992;6:100–9.
24. Steinberg D, Blank O, Rotstein I. Influence of dental biofilm on release of mercury from amalgam exposed to carbamide peroxide. J Biomed Mater Res B Appl Biomater. 2003;67(1):627–31.
25. Mahler DB, Pham BV, Adey JD. Corrosion sealing of amalgam restorations in vitro. Oper Dent. 2009;34(3):312–20.
26. Beyth N, et al. Streptococcus mutans biofilm changes surface-topography of resin composites. Dent Mater. 2008;24(6):732–6.
27. Bourbia M, et al. Cariogenic bacteria degrade dental resin composites and adhesives. J Dent Res. 2013;92(11):989–94.
28. Gregson KS, Shih H, Gregory RL. The impact of three strains of oral bacteria on the surface and mechanical properties of a dental resin material. Clin Oral Investig. 2012;16(4):1095–103.
29. Nedeljkovic I, et al. Biofilm-induced changes to the composite surface. J Dent. 2017;63:36–43.
30. Willershausen B, et al. The influence of oral bacteria on the surfaces of resin-based dental restorative materials—an in vitro study. Int Dent J. 1999;49(4):231–9.
31. Teughels W, et al. Effect of material characteristics and/or surface topography on biofilm development. Clin Oral Implants Res. 2006;17(Suppl 2):68–81.
32. de Paula AB, et al. Biodegradation and abrasive wear of nano restorative materials. Oper Dent. 2011;36(6):670–7.
33. de Gee AJ, et al. Influence of enzymes and plaque acids on in vitro wear of dental composites. Biomaterials. 1996;17(13):1327–32.
34. Hashimoto M, et al. In vivo degradation of resin-dentin bonds in humans over 1 to 3 years. J Dent Res. 2000;79(6):1385–91.
35. Kermanshahi S, et al. Biodegradation of resin-dentin interfaces increases bacterial microleakage. J Dent Res. 2010;89(9):996–1001.
36. Maske TT, et al. Minimal gap size and dentin wall lesion development next to resin composite in a microcosm biofilm model. Caries Res. 2017;51(5):475–81.
37. Thomas RZ, et al. Approximal secondary caries lesion progression, a 20-week in situ study. Caries Res. 2007;41(5):399–405.
38. Kuper NK, et al. Gap size and wall lesion development next to composite. J Dent Res. 2014;93(7 Suppl):108S–13S.
39. Li Y, et al. Degradation in the dentin-composite interface subjected to multi-species biofilm challenges. Acta Biomater. 2014;10(1):375–83.

40. Chrysanthakopoulos NA. Reasons for placement and replacement of resin-based composite restorations in Greece. J Dent Res Dent Clin Dent Prospects. 2011;5(3):87–93.
41. Burke FJ, et al. Influence of patient factors on age of restorations at failure and reasons for their placement and replacement. J Dent. 2001;29(5):317–24.
42. Mjor IA, Moorhead JE, Dahl JE. Reasons for replacement of restorations in permanent teeth in general dental practice. Int Dent J. 2000;50(6):361–6.
43. Asmussen E, Hansen EK. Surface discoloration of restorative resins in relation to surface softening and oral hygiene. Scand J Dent Res. 1986;94(2):174–7.
44. Ceci M, et al. Discoloration of different esthetic restorative materials: a spectrophotometric evaluation. Eur J Dent. 2017;11(2):149–56.
45. Hyun HK, Ferracane JL. Influence of biofilm formation on the optical properties of novel bioactive glass-containing composites. Dent Mater. 2016;32(9):1144–51.
46. Guler AU, et al. Effects of polishing procedures on color stability of composite resins. J Appl Oral Sci. 2009;17(2):108–12.
47. Borges MA, et al. Degradation of polymeric restorative materials subjected to a high caries challenge. Dent Mater. 2011;27(3):244–52.
48. Huang B, et al. Esterase from a cariogenic bacterium hydrolyzes dental resins. Acta Biomater. 2018;71:330–8.
49. Zhang SC, Kern M. The role of host-derived dentinal matrix metalloproteinases in reducing dentin bonding of resin adhesives. Int J Oral Sci. 2009;1(4):163–76.
50. Finer Y, Santerre JP. Influence of silanated filler content on the biodegradation of bisGMA/TEGDMA dental composite resins. J Biomed Mater Res A. 2007;81(1):75–84.
51. Finer Y, Santerre JP. The influence of resin chemistry on a dental composite's biodegradation. J Biomed Mater Res A. 2004;69(2):233–46.
52. Hagio M, et al. Degradation of methacrylate monomers in human saliva. Dent Mater J. 2006;25(2):241–6.
53. Gonzalez-Bonet A, et al. Preparation of dental resins resistant to enzymatic and hydrolytic degradation in oral environments. Biomacromolecules. 2015;16(10):3381–8.
54. Decha N, et al. Synthesis and characterization of new hydrolytic-resistant dental resin adhesive monomer HMTAF. Des Monomers Polym. 2019;22(1):106–13.

8 Considerations for Designing Next-Generation Composite Dental Materials

Carmem S. Pfeifer, Jens Kreth, Dipankar Koley, and Jack L. Ferracane

Abstract

Tooth-colored dental restorations, especially the ones produced chairside, have relatively short durability, leading to additional tooth loss and high costs. This chapter reviews previous research and highlights future considerations for designing new dental composite restorative materials. Materials that address failure due to the interactions with oral biofilms are explored, since secondary caries is a leading cause for restoration replacement. The process of dental caries is briefly reviewed, emphasizing the surface interactions between biofilms and materials, and the tooth structure. Current research into materials design solutions is described, and future perspectives are discussed. Importantly, a novel method for studying ion release from new dental materials and their interaction with the oral biofilm is presented.

C. S. Pfeifer (✉) · J. Kreth · J. L. Ferracane
Department of Restorative Dentistry, Oregon Health & Science University, Portland, OR, USA
e-mail: pfeiferc@ohsu.edu; kreth@ohsu.edu; ferracan@ohsu.edu

D. Koley
Department of Chemistry, Oregon State University, Corvallis, OR, USA
e-mail: Dipankar.Koley@oregonstate.edu

8.1 Introduction

The primary reasons that resin-based dental composite restorations of posterior teeth are considered to have failed and need replacement are material fracture and the diagnosis of recurrent caries [1, 2], the latter being most typically associated with the gingival margins of interproximal restorations [3]. This outcome suggests that the clinical performance of dental composite restorations could be improved if the material were specifically designed to make the restored tooth more resistant to microbial biofilms and/or their secreted metabolic products. Attempts to do this have addressed the likely material qualities that influence this process, such as resin shrinkage during the polymerization process that creates substantial stress and leads to marginal debonding and gap formation, improved dental adhesives to resist resin contraction forces, and enhanced adhesive/composite formulations that resist intraoral degradation, especially at the marginal interface, in part by the elimination or reduction of ester groups in the monomers.

In order to design new materials that can resist the deleterious effects produced by biofilms, it is important to understand the conditions driving dental biofilm formation at or near restoration margins. The oral cavity contains a rich and diverse popula-

A. C. Ionescu, S. Hahnel (eds.), *Oral Biofilms and Modern Dental Materials*, https://doi.org/10.1007/978-3-030-67388-8_8

tion of microbes, and particularly important are acidogenic species that cause the demineralization of adjacent tooth structures. Understanding the mechanism by which these microbes adhere to tooth surfaces and dental restorative materials to form complex and potentially antibiotic-resistant biofilms is crucial for determining how to deter them. Further, it is critical to determine what potential components of dental composites can affect biofilm adhesion, viability, and virulence, and then whether it is possible to include these components into a material without negatively affecting its properties and performance. Many examples of the incorporation of potentially antimicrobial components have been proposed [4, 5]. The most examined approaches have been the development of materials capable of releasing specific ions, such as silver or other metals, that have microbe-killing potential, as well as the incorporation of organic antimicrobial compounds, such as quaternary ammonium molecules, which may be either released to kill microbes or bound to serve as contact-killing agents [6]. More in-depth information on this topic is available in Chap. 10. To date, no successful solution has been clinically proven, which explains why the main cause of failure for these materials has not changed for decades.

The overall goal of this chapter is to review previous research and highlight future considerations for designing new dental composite restorative materials, specifically those that can address failure due to the recurrence of dental caries. To accomplish this, the process of dental caries is reviewed, with specific emphasis on the potential interaction between microbial biofilms and tooth and dental composite surfaces. The materials design solutions currently being studied to address this issue are then discussed, as well as future ideas for producing improved materials. Finally, the development of a novel method for studying ion release from new dental materials and their interaction with the oral biofilm is presented.

8.2 Section 1: Dental Biofilm Formation

Dental biofilm formation follows distinctive temporal and spatial patterns. Recent advances in oral microbiome composition studies using next-generation sequencing techniques have identified a highly diverse microbiota present on teeth with considerable variation among different subjects [7–9] and Chap. 1. At the same time those microbiome studies have challenged oral disease etiologies. For example, the archetypic caries pathogen *Streptococcus mutans* is rarely found in deeper caries lesions and its abundance can be relatively low in other caries lesions [9–11] and Chap. 3. Furthermore, advanced imaging techniques used to visualize dental plaque biofilm formation in situ [12] have also challenged the traditional picture of biofilm development based on in vitro co-aggregation studies [13, 14]. Yet those new developments are rarely considered as basis for the development of new dental materials. Overall biofilm formation is highly complex and therefore difficult to recreate in the laboratory setting. However, careful planning can include some of the new developments in experimental setups to develop and test new dental materials. This section describes some of the concepts of biofilm formation and discusses its implications on dental material development. For a more detailed description on molecular aspects of dental biofilm formation see [15].

8.2.1 It Started with a Film

Initial biofilm formation requires attachment of bacterial cells to the tooth surface. The pioneer colonizer group of streptococci is able to directly adhere to the major mineral found in dental enamel, hydroxyapatite [16]. However, in vivo the tooth and any dental composite are covered with a saliva-derived film called the acquired enamel pellicle (AEP) [17]. Its importance and main roles are described in Chap. 2. Streptococci are especially well equipped with adhesins that recognize salivary pellicle proteins. One of the most abundant proteins in AEP is α-amylase, catalyzing starch hydrolysis, and the ability to directly interact with amylase is conserved among streptococci [18]. Furthermore, the major mucins MUC7 (low molecular weight) and MUC5B (high molecular weight) are prevalent in AEP but differ in abundance [19]. Mucins, produced by salivary glands, are glycoproteins and form a lubricating, viscoelastic coating on all oral surfaces and pioneer colonizers are able to bind to mucins [20]. Why are amylase and mucins relevant in the context of dental material development? Both proteins determine selectivity for specific bacterial groups and amylase and MUC5B seem to be depleted in the AEP compared to saliva, but this was not seen for MUC7 [19]. Interestingly, one of the most abundant streptococcal species associated with oral health, *Streptococcus sanguinis* shows a high specificity for binding MUC7 via surface protein SrpA [21]. Conceptually, developing dental materials that support binding of MUC7, which predominates in saliva of caries-resistant individuals [22], and at the same time deterring MUC5B and amylase binding would not only promote attachment of a health-associated species, able to delay colonization of cariogenic species like *S. mutans* [23], but also help in the clearance of other species that bind to MUC5B and amylase through saliva flow [24]. Moreover, divalent cations like Ca^{2+} and Mg^{2+} present in saliva in concentrations greater than 1 mM in healthy individuals [25] play an important role in the binding ability of oral streptococci to the AEP. Chelation of both cations decreased binding to several salivary and AEP components [26]; thus in any biofilm experiment testing dental materials, those cations should be present, ideally in saliva. These two examples illustrate the importance of considering how the biological environment can influence binding and community selection of the oral biofilm.

8.2.2 Biogeography: Where to Attach Matters

After initial attachment to the AEP, biofilm development is defined in several stages, which can be distinguished genetically [27]. The dental biofilm develops into a dynamic multispecies population (including bacteria, viruses, fungi, archaea, and certain protozoa) comprised of up to 30,000 species in any one individual, with an overall human population-wide microbial richness that has not been fully assessed. Once initial attachment has been successful, which is a reversible process, bacteria fully commit to the biofilm lifestyle, which includes the formation of the biofilm-stabilizing and -protecting extracellular polysaccharide matrix (EPS) (Fig. 8.1). First, microcolonies of mainly *Streptococcus* and *Actinomyces* start to develop on the tooth surface and restorative materials by cell proliferation and integration of free-floating microbes from saliva [28]. The biofilm matures over time and the last stage includes a dispersal process to colonize new sides (Fig. 8.1). However, a recent study shows that the process is much more elaborate and that a previously underestimated species plays a significant role in supragingival plaque formation. In this study, *Corynebacterium* was established as a key taxon in supragingival plaque. *Corynebacterium* seems to adhere at the gingival margin forming spatial structures that resemble hedgehogs, with several other species, including streptococci attaching to the long filaments *Corynebacterium* forms [12]. The key in this observation lies in the fact that mature biofilms in vivo have a spatial arrangement and species composition that single or multispecies in vitro systems cannot reconstitute.

The question that arises is if single or multispecies biofilms comprised of a few species are relevant and adequate in material testing. For example, a common bacterial species used is *S. mutans* due to its ability to form very tenacious and "sticky" EPS made from sucrose. In addition, *S. mutans* serves as a model for lactic acid production, which is responsible for tooth demineralization and caries development [29]. Overall *S. mutans* seems like a logical candidate. However, as of today it is not known where *S. mutans* spatially organizes in the dental biofilm. There is no doubt that it is frequently isolated from subjects with caries, but is the majority of *S. mutans* in direct contact with the AEP and therefore potentially in contact with the dental material? Or is *S. mutans* associated with other bacterial species in structures that are located close to the tooth surface but not in direct contact, and thus contact-dependent killing mechanism would not be very effective? Similar, does the EPS provide a "buffer" for *S. mutans* that shields it from direct contact with the AEP and dental material?

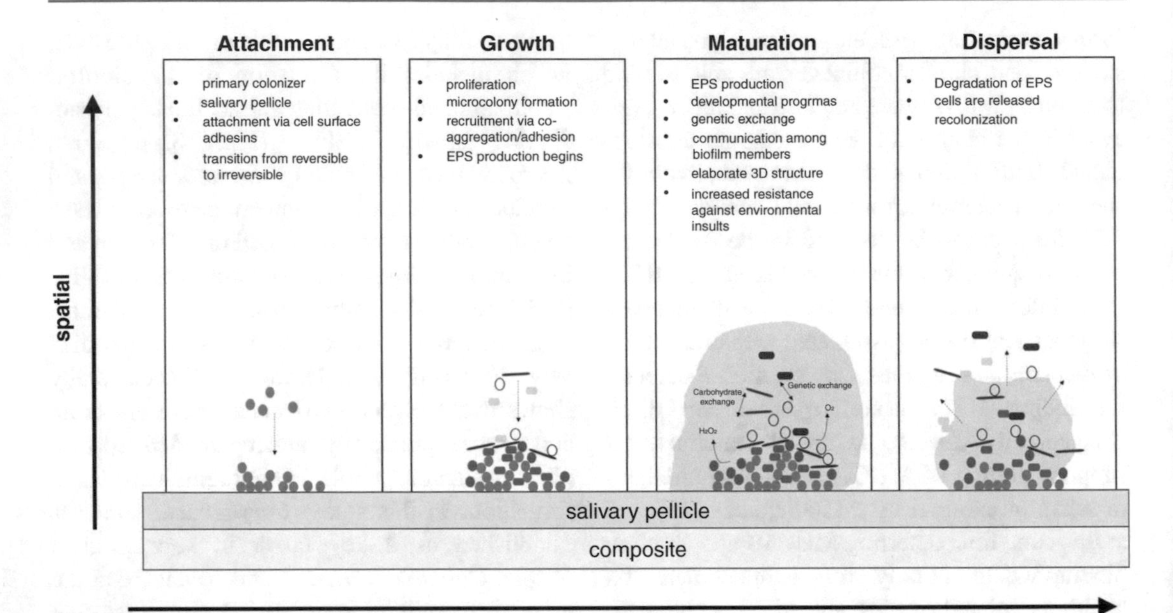

Fig. 8.1 Adaptation of identified stages of biofilm formation to the dental composite surface. During biofilm formation several stages occur. During the attachment stage, early colonizing species attach to the proteinaceous salivary pellicle (AEP) on the tooth enamel via adhesins. Secondary and late colonizers will co-aggregate using receptors present on the early bacteria and each other to build the biofilm during the growth stage. For the maturation stage, bacteria begin to communicate metabolically through the release of small molecules and substrates for cross feeding. Antimicrobial components such as H_2O_2 and bacteriocins are produced increasing competition. Additionally, genetic information is exchanged via the release of extracellular DNA, which also plays a structural role in the biofilm. Green circles represent early colonizers; orange circles and blue rods, secondary colonizers; purple rods, fusobacteria; white circles and yellow and red rods, late colonizers. In the third and fourth panels, the grey represents the exopolysaccharide matrix (EPS)

The difference of *S. mutans* isolated from distinct intraoral sites or subjects further complicates the selection of appropriate *S. mutans* strains. Some of the strains are good biofilm formers and others have a better ability to withstand stress [30].

One of the stunning observations of biofilm architecture is the development of water channels, which enables the biofilm community to exchange metabolites and signaling molecules and dispose of toxic compounds that otherwise would accumulate in the biofilm structure [31]. *S. mutans* can be used again as an example, since it forms in vitro biofilms in the presence of sucrose that are very compact, lacking comparable structures developed by in situ dental plaque [12, 32]. This allows *S. mutans* to limit diffusion of lactic acid. However, antimicrobial components leaching into *S. mutans* in in vitro biofilms could reach potentially higher local concentrations due to the diffusion limitations and exert antimicrobial activity, while in vivo the local concentration might be much lower due to the water channels and salivary flow.

For material testing, the position and type of the colonizing microbe, antimicrobial susceptibility, or general ability to cope with stress is very relevant and should be modeled as close to the in vivo situation as possible.

8.2.3 Biofilms: The Smart Ones

Microbes in general are perfectionists when it comes to adaptation. There are several reasons for this, one of them being the ability to exchange genetic material [33]. It has been shown that dental biofilm bacteria, especially the group of oral streptococci, are able to develop a physiological state called competence that enables the uptake of extracellular DNA present in biofilms [34]. This developmental state is induced under biofilm growth conditions and allows among other things for the exchange of antibiotic resistance genes. Considering that the trend is towards the development of dental materials with antimicrobial or antifouling activity, it is surprising that the development of resistance mechanisms is seldom addressed. In vitro development of biofilms on dental materials might not be reflective of the in vivo situation regarding the development of resistance mechanism. Any dental material in vivo is always adjacent to a sound tooth surface, which allows unaltered biofilm formation. Therefore the biofilm might come into close proximity with the dental material, just close enough to be exposed to the antimicrobial mechanism of the material without getting severely affected. This is the ideal scenario for the development of resistant mechanisms.

8.2.4 Biofilms on Dental Materials: What Are We Missing?

Recreation of complex biological systems is not always feasible and requires reductionist approaches. Nonetheless, information about dental biofilm formation in situ, species composition, and microbial metabolic activity, as well as the oral microbial secretome, is available [12, 35, 36]. A careful assessment of the biological information and necessary implementation of this knowledge into dental material development and testing are possible. Unfortunately, key information about certain aspects is still missing. How biofilm architecture and development appear on dental materials in subjects is not known. Similar, whether or not the biofilm species composition on dental materials resembles the composition on sound dental surfaces needs to be determined. The same holds true for AEP.

8.3 Section 2: Formulation of Materials with Biological Activity

Intense research has developed in the area of antimicrobial and antifouling materials for direct restorative applications [37], with the main objective of impeding or at least hampering the attachment of caries-forming bacteria with subsequent biofilm maturation. These materials and compounds will be summarized in the first part of this section. If biofilm development cannot be prevented, a possible second line of defense is to remineralize the lost tooth structure. This will be the focus of the second part of this section.

8.3.1 Preventing Bacterial Attachment and Biofilm Formation: Antibacterial Compounds Applied to Dental Composites

Antibacterial materials have been developed to target different aspects of biofilm establishment, including direct bacterial destruction via small molecule release [38] or contact kill mechanisms [6], antifouling strategies [39], and disruption of the extracellular matrix [40]. Classic examples of small molecule release include silver particles, triclosan- and chlorhexidine-containing compounds [41], which when incorporated directly into the matrix result in burst release but not a sustainable antibacterial effect [38]. In contrast, the incorporation of antimicrobial agents within mesoporous silica particles allows for controlled release over longer periods of time [38]. However, these have had limited use due to concerns over deteriorating mechanical and optical properties resulting from water exchange.

An alternative to incorporating leachable compounds is to produce molecules that can be covalently attached to the polymer network of the restorative materials, such as polymerizable quaternary ammonium methacrylates [42, 43]. These materials have a quaternized, cationic nitrogen atom associated with a side alkyl chain of varied length (Fig. 8.2a). The proposed mechanism of action involves attraction of the negatively charged bacterial wall by the positive charge in the monomer. Once this contact is established, the long alkyl chain interacts with the lipoproteic membrane and "bursts" the bacterial wall [44]. The length of the alkyl side chain is critical to this mechanism [45], with one study showing a five-fold reduction in bacteria when comparing chains

a General structure of a QAM b 2-methacryloyloxyethyl phosphorylcholine

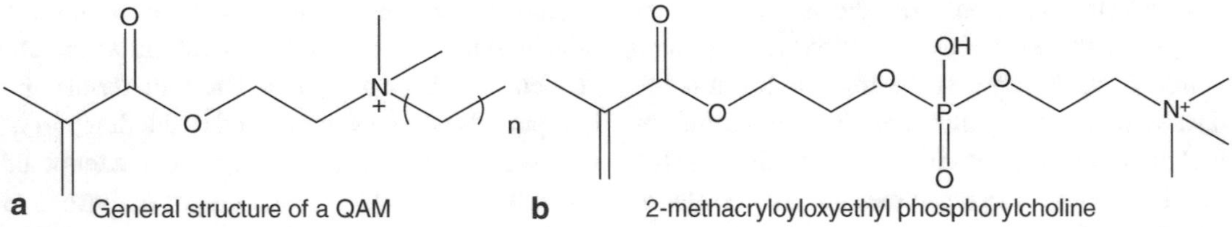

Fig. 8.2 (**a**) General structure of a quaternary ammonium methacrylate. The size of the side chain (*n*) has been correlated with the antibacterial activity. (**b**) 2-Methacryloyloxyethyl phosphorylcholine—MPC. This monomer has been added to composites to impart antifouling properties

with 16 vs. 3 carbons [44]. The antibacterial effect is also a function of the surface charge concentration [46]. For some compositions, positive effects are achieved without compromising mechanical properties [47]. A common criticism to this approach is that it requires surface contact, and even if the bacteria are killed upon contact, disrupting the initial formation of a biofilm, if the early colonizers are not effectively removed from the surface, subsequent colonization onto the debris left behind still can occur with formation of a potentially virulent biofilm. While in a few studies QAMs showed effectiveness in the bulk of mature biofilms [48], the possibility that this was related to leaching of antimicrobial monomers that did not properly polymerize cannot be eliminated, also creating concerns over cytotoxicity and overall material stability [49]. Moreover, the effect of QAMs on persister cells (dormant bacterial cells highly resistant against antimicrobial killing) has not been completely elucidated [50]. Despite intense research, there is only one example of commercial QAM-based material (Clearfil Protect Bond, Kuraray).

QAMs have also been proposed as antifouling monomers, such as 2-methacryloyloxyethyl phosphorylcholine (MPC, shown in Fig. 8.2b) [51, 52]. This polymerizable methacrylate contains a phospholipid chain that reduces protein absorption, bacterial adhesion, and cellular attachment to reduce pellicle formation [51, 52], and has been shown to lead to a fourfold decrease in biofilm formation compared to unmodified controls, acting synergistically with other QAMs added to the composition [52].

More recently, antibacterial materials based on zwitterionic compounds (carboxybetaine and sulfobetaine structures) have been investigated (Fig. 8.2a). These materials are responsive to environmental conditions such as pH and hydration [53, 54]. One carboxybetaine compound was shown to reversibly switch between an open-ring, antifouling conformation and a closed-ring, antibacterial conformation as a function of the pH, and significantly decreased *E. coli* biofilms [54]. With this mechanism (Fig. 8.2b), the initial goal is to impede bacterial attachment to the surface, but if this does not happen, once the formed biofilm creates acid and the pH is reduced, the molecule switches and becomes bactericidal [54]. These compounds, however, also rely on a contact kill mechanism, and their efficacy against the *Streptococcus* genus, which is knowingly more resilient than *E. coli*, has not been investigated to date.

Functionalized diamond nanoparticles have been demonstrated to reduce the formation of some biofilm species, such as *E. coli*, but were not effective against *S. aureus* [55]. Glycan functionalization in particular seems to be effective against *E. coli* [56], while oxidized and negatively charged surfaces have shown antibacterial properties against *E. coli* and *B. subtilis* [57]. Untreated particles did not seem to prevent Pseudomonas biofilm growth, regardless of particle size and distribution [58]. These particles have yet to be tested against the more resilient *Streptococcus* genus.

More recently, materials targeting the disruption of the formation of the biofilm extracellular matrix have been investigated principally concentrating on compounds capable of disrupting the formation of EPS [40]. Glucosyl transferase (GTF) is a significant enzyme in this process. A significant amount of small-molecule GTF inhibitors have been screened, including

2-aminoimidazole and its derivatives [59], hydrochalcones [60], the FDA-approved anticancer and antimicrobial drug trimetrexate [61], and several others specifically designed using in silico docking to selectively target the active site of the catalytic domain from *S. mutans* GTF [40]. The inhibition mechanism varies (either direct downregulation of gene expression or inhibition of GTF binding—[40]) or regulates cell growth. All of these compounds have the advantage of being highly selective to pathogenic bacteria (*S. mutans*), but not commensals (i.e., *S. sanguinis*), and being effective at relatively low concentrations (low MIC values) [40], though their release must be controlled to have a sustained effect. Efforts are underway to make similar molecules polymerizable, such as imidazolium dimethacrylate, which has been shown to disrupt biofilm formation when added at very low concentrations (2%) to a dimethacrylate monomer matrix of a filled composite, without compromise to the short-term mechanical properties [62]. Finally, specifically targeted antimicrobial peptides have been shown to selectively kill *S. mutans* with high efficacy, while at the same time favoring the colonization of commensal streptococci [63].

Other strategies, such as the use of surfaces with super-hydrophobic and super-hydrophilic character [64], or with micro- or nano-patterned structures [65], are also promising for use in biomedical applications including dentistry. For example, one study has demonstrated that variations in topographical features, such as nanopattern size, concentration, and spatial distribution, can significantly impact the adhering pattern and stretching degree of bacterial cell membranes [66]. Others have demonstrated increased antibacterial effect when nanopatterned surfaces were combined with antimicrobial peptides [67]. These strategies represent a safer alternative to antibiotic-derived therapies.

8.3.2 Remineralizing Materials

There is a movement in dentistry to take a more conservative, nonsurgical, approach to restorative dentistry, with emphasis on the remineralization of non-cavitated lesions and tooth structure associated with restorations (see also Chap. 9). The most common available materials are based on fluoride-releasing formulations, including glass ionomer cements, which have demonstrated some success at preventing secondary caries formation via displacement of the de/remineralization equilibrium towards the remineralization side [68]. Fluoride also promotes the formation of the less soluble fluorapatite [69], and has some direct antimicrobial effect [70]. However, this approach is only effective in small, incipient lesions and requires the release of fluoride ions into the oral cavity [71]. Highly filled fluoride-releasing composites have shown significantly lower fluoride release than glass ionomers [72], and the fluoride addition may reduce mechanical properties [73]. One experimental material based on novel chelating monomers demonstrated sustainable fluoride release with mechanical properties similar to conventional composites [74], but the remineralization capability was not assessed. Others have attempted to associate amorphous calcium phosphate (ACP) in one fluoride-releasing formulation, rendering better remineralization potential without affecting mechanical properties [75].

While it is intuitive to attempt to use crystalline calcium phosphates for remineralization in vivo, this is difficult due to the inherent low solubility of the compounds, especially in the presence of fluoride ions [76]. Brushite and β-tricalcium phosphate (β-TCP) are the most soluble among the crystalline phases, and therefore the more heavily studied, demonstrating good ion release from brushite-containing dental composites for selected formulations [77] and for β-TCP functionalized with sodium lauryl sulfate [78]. Their remineralization potential has yet to be investigated. Crystalline calcium phosphates have also been incorporated in sol-gel-processed or melt-derived bioactive glasses [79, 80], with some evidence for remineralization of surrounding tooth structure [81], and desensitization via dentin tubular occlusion [82]. More recently, a bioactive glass produced via a sol-gel mechanism has been shown to decrease bacterial colonization and demineralization in restoration marginal gaps in a secondary caries model [83].

Composites formulated with amorphous calcium phosphates (ACP) produce enamel remineralization around restorations, but high concentrations of ACP reduce mechanical properties [84]. When the ACP is limited to around 10%, remineralization effects persist without a loss of mechanical properties [85], especially if the ACP particles are silanized [86]. Up to 40 wt% ACP has also been added to dental adhesives, where the mechanical requirements are not as stringent, and calcium and phosphate ion release at low pH (~4) has been demonstrated without compromising microtensile bond strength [87]. Some studies have shown that the incorporation of nanocomplexes of carboxymethyl chitosan/amorphous calcium phosphate is capable of mimicking the stabilizing effect of dentin matrix protein 1 (DMP1) on ACP, which helps guide dentin remineralization in a more organized, hierarchical fashion [88]. ACP has also been combined with quaternary ammonium methacrylate monomers [42, 89] in composites and adhesives to render the material antimicrobial and remineralizing at the same time. The combination of QAM monomers and ACP in composites did not affect properties, and produced in vitro antimicrobial effects against various oral pathogens and remineralization [90].

Calcium silicates are being used as remineralizing agents, through calcium release as well as via stimulation of transforming growth factor (TGF) beta-1 production, which in turn leads to progenitor cell differentiation and dentin tissue formation at the interface with the material [91]. These materials have poor mechanical proper-

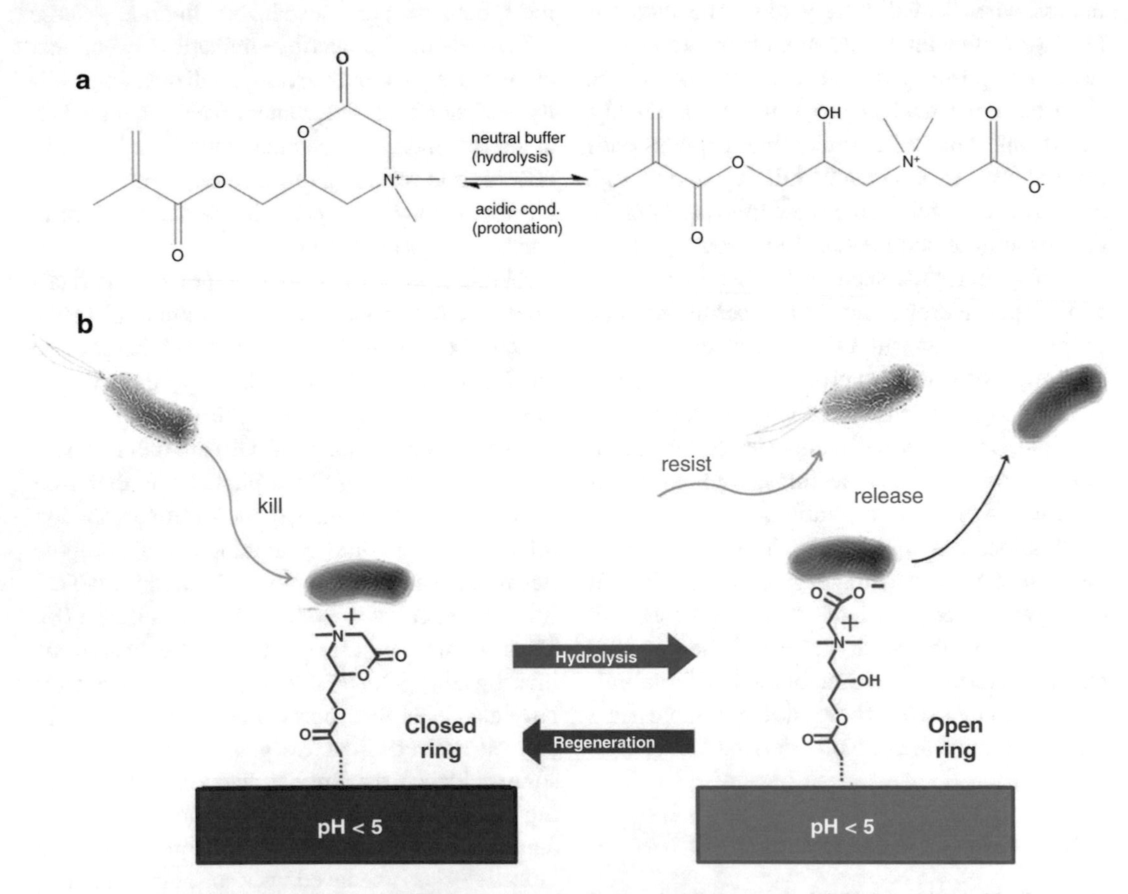

Fig. 8.3 (**a**) General structure of a zwitterionic methacrylate in the closed- and open-ring conformations. The monomer switches between these two conformations according to the conditions of the environment, and alternates between antimicrobial and antifouling. (**b**) Schematic representation of the antimicrobial vs. antifouling switching of zwitterionic surfaces. (Modified from [54])

ties, but products have been commercialized for pulp capping applications (Theracal®, Biodentine®, Activa®). Biodentine presented similar clinical performance to mineral trioxide aggregate (MTA, or Portland cement) [92], and has been further modified to include fluoride-release capabilities [93], though the anticaries potential has yet to be investigated. Finally, high-pH materials designed for pulp capping applications, based on either calcium hydroxide or MTA, have demonstrated consistent remineralization potential in deep dentin lesions and are widely available commercially [94]. More recent developments, such as modification with photoactivated resins, have aimed at improving their mechanical strength and stability [95] (Fig. 8.3).

8.4 Section 3: Novel Method for Characterizing New Dental Material Formulations

As materials become more highly developed, so too does the need for technology that can be used to characterize them. In addition to establishing the structure and physical properties of these new materials by use of microscopy and mechanical based techniques, there is a growing need to understand the chemical properties of these new composites, such as the amount of metal ions or other polyanions that they release. These parameters affect the longevity of the materials and their efficacy in preventing biofilm growth on their surfaces. At present, the most commonly used techniques involve collecting an aliquot from an overnight composite soaking solution and injecting it into an inductively coupled plasma mass spectrometer to quantify the amount of metal ions. However, the major shortcomings of these techniques are that they do not provide any information about the real-time release pattern, especially the local chemical environment just above the composites. The local environment is especially important for designing composites that can control biofilm formation, as bacteria are able to sense only a couple of hundreds of micrometers above the composites. To address these shortcomings, a new electrochemical sensor-based analytical technique has been developed, scanning electrochemical microscopy (SECM), in which particular sensors (pH, Ca^{2+}, H_2O_2, etc.) are used as chemical probes to map the corresponding parameter in three-dimensional space above the composites [96–101].

8.4.1 Introduction of SECM

The basic operating principles of SECM have been described [102–104], and a full in-depth technical description of this technique can be found in the monograph by Bard and Mirkin [105]. SECM is a nondestructive scanning probe technique used to characterize both soft and hard surfaces, such as live biofilm or dental composites. A unique feature of this technique is the ability to position the chemical probe at a known distance from the substrate (biofilm or composite) without touching it. In SECM, the probe is positioned by using a distance-current calibration curve or feedback approach curve (as it is commonly known in the SECM literature). Typically, a 25 μm Pt electrode or tip is mounted on a high-resolution *x–y–z* stepper motor while the substrate is placed inside a solution-containing petri dish on the SECM stage, as shown in Fig. 8.4a. A reference (Ag/AgCl) and a counter electrode (0.5 mm Pt wire) are also placed in the same solution to complete the electrical connection with the working electrode or the 25 μm Pt electrode/tip. In practice, potassium ferrocyanide (because of its nontoxicity towards bacteria) or oxygen is added to the solution and a diffusion-controlled potential (a high overpotential to oxidize or reduce everything at the electrode surface) is applied to the tip while it simultaneously moves in the *z*-direction towards the substrate. As the tip approaches an insulating substrate, the diffusion of the redox molecules present in the solution becomes blocked by the insulating glass sheath surrounding (Fig. 8.4a) the tip, causing the current to drop, as shown in the negative feed-

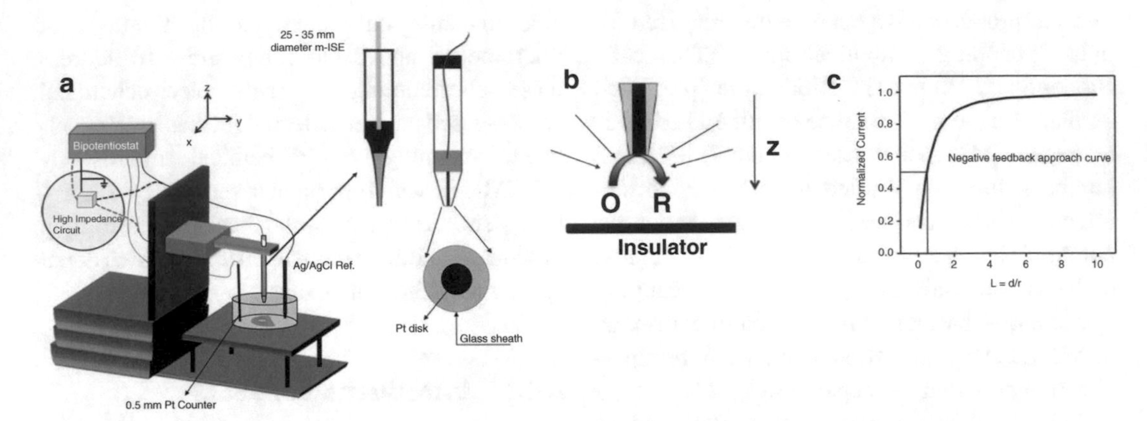

Fig. 8.4 (**a**) Schematic diagram of the scanning electrochemical microscope (SECM). The arrow showed the two different types of electrodes used as a SECM chemical probe. The 25 μm Pt electrode is generally being used as an amperometric mode redox sensor and the 25–35 μm electrode is a specially designed dual-function (amperometric-potentiometric) mode sensor to map redox molecules and cations, respectively. (**b**) The schematic representation of the blocking of diffusion of oxidized species (O) towards the probe where the O is reduced instantly at the electrode surface. O and R represent the oxidized and the reduced forms of the same redox molecule. (**c**) The current-distance response curve or negative feedback approach curve used to position the probe. The *x*-axis is the normalized distance where *r* is the radius of the probe and *d* is the distance between the probe and the substrate. The *y*-axis is normalized current where the current is normalized by the current recorded by the probe at $L = 10$. The approach curve is universal current-distance working curve as it is valid for any redox molecules and any probe size

back approach curve (Fig. 8.4b, c). In contrast, when the tip approaches a conducting surface, where the redox species produced by the probe is regenerated, the redox probe measures an increase in current and a positive feedback approach curve is obtained. In biological or composite testing samples, however, the positive feedback approach is rarely observed or used. The approach curve is plotted as a normalized current versus a normalized *z*-direction distance (see Fig. 8.4). After the distance between the probe and substrate is fixed, the solution is replaced with the relevant buffer solution without the redox mediator and the SECM probe is switched to the desired potential to detect and quantify the molecule of interest. The probe can also be switched to a potentiometric mode in which the potentials generated by the ions of interest can be measured with respect to time. For example, pH and Ca^{2+} can be measured selectively and quantitatively by using a SECM potentiometric probe, as discussed later.

8.4.2 Characterization of Biomaterials/Biofilm Interactions with SECM

New types of solid-state ion-selective electrodes have been developed for use as a SECM chemical probe to characterize the local pH and Ca^{2+} released by ion-releasing composites, such as those containing a bioactive glass (BAG) (Fig. 8.5a). One of the major limitations in miniaturizing the traditional plastic-based ion-sensing membrane is its insulating nature in conducting current, as well as its high impedance. Thus, our lab designed two individual sensors: (1) the first is a dual-SECM probe, with one 25 μm Pt electrode to perform the amperometric approach curve and fix the probe distance and another 25 μm Pt electrode coated with polyaniline to act as a pH sensor; this dual probe is used to quantitatively map the local pH change caused by the BAG composite substrate (Fig. 8.5a, b) when exposed to pH 4.5 saliva; (2) the second type of

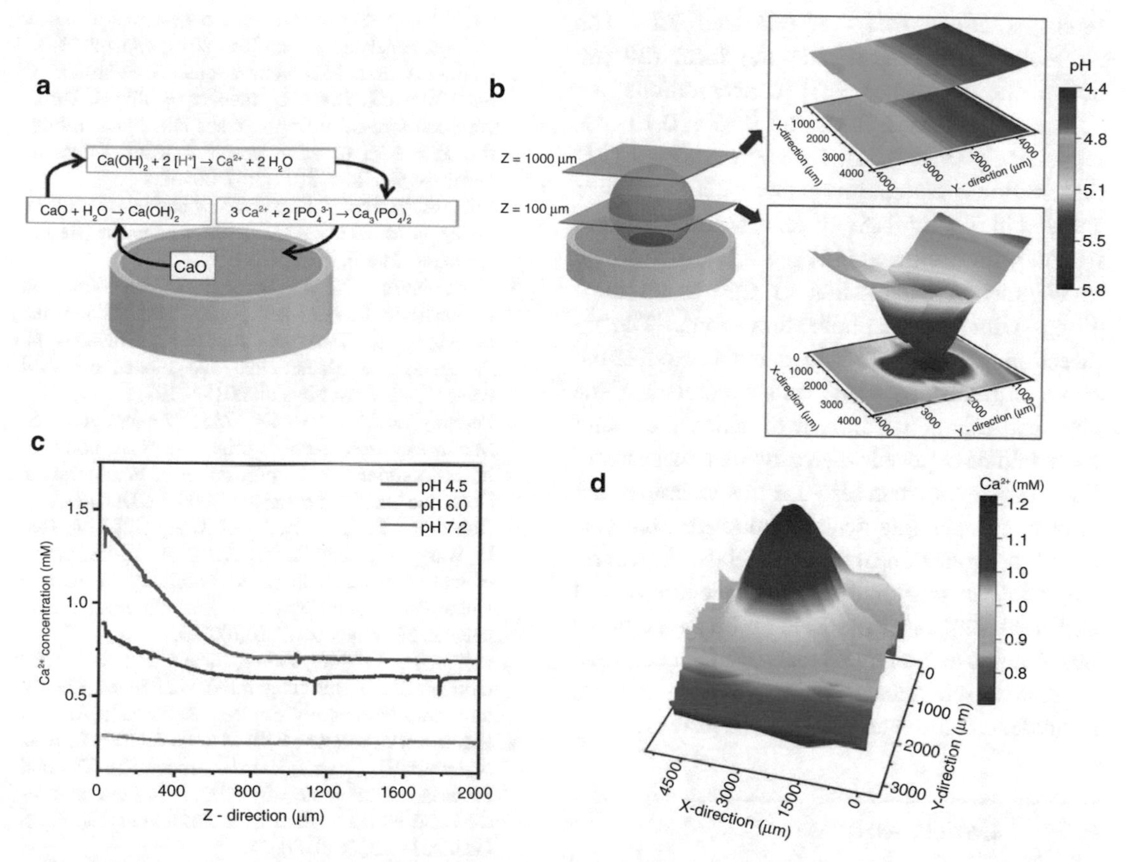

Fig. 8.5 (**a**) Schematic diagram of pH-dependent calcium ion release from pure bioactive glass (BAG) when immersed in acidic solutions (pH 5.5 or lower). (**b**) Three-dimensional SECM image of pH distribution above the BAG surface. pH imaging was performed over 1.6 mm of the exposed BAG surface (100 and 1000 μm above) in artificial saliva at pH 4.5. The transparent purple sphere represents the neutralized zone produced by the BAG. (**c**) *z*-Direction Ca^{2+} profile above the BAG substrate in the presence of artificial saliva of pH 4.5, 6.0, and 7.2. (**d**) Three-dimensional SECM image of calcium ion release from the BAG surface when exposed to artificial saliva (pH 4.5). The image was taken at a constant height of 200 μm from the BAG surface

SECM probe is a carbon-based, calcium ion-selective microelectrode (Ca^{2+}-μISME), 25 μm in diameter, used to correlate the change in local pH and the corresponding Ca^{2+} release profile above the BAG composite surface. This unique Ca^{2+}-μISME (ion-selective microelectrode) is capable of performing an amperometric approach curve and serving as a potentiometric Ca^{2+} sensor. These sensors show a lower detection limit of 1 μM and broad linear working range of 5 μM–200 mM with a near-Nernstian slope of 28 mV/log $a_{Ca^{2+}}$. The response time of these sensors is around ~200 ms while being highly selective against major interfering ions. The selectivity coefficients of the Ca^{2+}-ISMEs are log $K_{Ca^{2+},A}$ = −5.88, −5.54, and −6.31 for Mg^{2+}, Na^+, and K^+, respectively. Due to the high carbon content in these Ca^{2+}-μISME, it is capable of performing an amperometric approach curve that aids the probe in being positioned at a known distance above the substrate. This allows the probe to quantitatively map the Ca^{2+} microenvironment produced by a pure BAG substrate in the presence of three different solutions (pH 4.5, 6.0, and 7.2). The z-direction scan shows that the local (20 μm above the substrate) Ca^{2+} concentrations are 1.40 ± 0.10, 0.85 ± 0.05, and 0.35 ± 0.10 μM, respectively (Fig. 8.5c). Three-dimensional Ca^{2+} distribution chemical image at z = 200 μm is also showed in Fig. 8.5d. As evident from the concentration profile showed in Fig. 8.2c, d, the change in chemical concentration of Ca^{2+} is all local along with the neutralization zone. The z-direction pH profile mapped the neutralization zone produced by BAG and to elucidate the chemical microenvironment to which the bacteria would be exposed when grown on these materials. Hence, one can also use this technique not only to characterize dental composites, but also to aid in the design of new materials. For example, one can design different concentrations of Ca^{2+}-releasing BAG particles and resin composites to achieve the target local Ca^{2+} concentration to create the chemical microenvironment required to hinder or eliminate the bacterial growth.

8.5 Conclusion

It is logical to assume that future dental composite development will be aimed at making materials that are antimicrobial, as well as having the capacity to remineralize adjacent tooth structure. Understanding the formation and function of oral biofilms is critical for accomplishing this goal. With this knowledge, the addition of appropriate specific compounds to the resin, either as leachable or bound organic compounds or as ion-releasing fillers, can be attempted to produce effective materials. Further development and use of highly sensitive sensing methods, such as SECM, are important in the design process, by providing an accurate assessment of the temporal and spatial release of these agents under appropriate conditions.

References

1. Opdam NJ, van de Sande FH, Bronkhorst E, Cenci MS, Bottenberg P, Pallesen U, Gaengler P, Lindberg A, Huysmans MC, van Dijken JW. Longevity of posterior composite restorations: a systematic review and meta-analysis. J Dent Res. 2014;93(10):943–9.
2. Rasines Alcaraz MG, Veitz-Keenan A, Sahrmann P, Schmidlin PR, Davis D, Iheozor-Ejiofor Z. Direct composite resin fillings versus amalgam fillings for permanent or adult posterior teeth. Cochrane Database Syst Rev. 2014;3:CD005620.
3. Mjor IA, Toffenetti F. Secondary caries: a literature review with case reports. Quintessence Int (Berlin, Germany: 1985). 2000;31(3):165–79.
4. Chatzistavrou X, Lefkelidou A, Papadopoulou L, Pavlidou E, Paraskevopoulos KM, Fenno JC, Flannagan S, Gonzalez-Cabezas C, Kotsanos N, Papagerakis P. Bactericidal and bioactive dental composites. Front Physiol. 2018;9:103.
5. Pereira-Cenci T, Cenci MS, Fedorowicz Z, Marchesan MA. Antibacterial agents in composite restorations for the prevention of dental caries. Cochrane Database Syst Rev. 2009;3:CD007819.
6. Han Q, Li B, Zhou X, Ge Y, Wang S, Li M, Ren B, Wang H, Zhang K, Xu HHK, et al. Anti-caries effects of dental adhesives containing quaternary ammonium methacrylates with different chain lengths. Materials. 2017;10(6):643.
7. Eriksson L, Lif Holgerson P, Johansson I. Saliva and tooth biofilm bacterial microbiota in adolescents in a low caries community. Sci Rep. 2017;7(1):5861.
8. Ribeiro AA, Azcarate-Peril MA, Cadenas MB, Butz N, Paster BJ, Chen T, Bair E, Arnold RR. The oral bacterial microbiome of occlusal surfaces in children and its association with diet and caries. PLoS One. 2017;12(7):e0180621.
9. Simon-Soro A, Guillen-Navarro M, Mira A. Metatranscriptomics reveals overall active bacterial composition in caries lesions. J Oral Microbiol. 2014;6:25443.
10. Mira A. Oral microbiome studies: potential diagnostic and therapeutic implications. Adv Dent Res. 2018;29(1):71–7.
11. Rocas IN, Alves FR, Rachid CT, Lima KC, Assuncao IV, Gomes PN, Siqueira JF Jr. Microbiome of deep dentinal caries lesions in teeth with symptomatic irreversible pulpitis. PLoS One. 2016;11(5):e0154653.
12. Mark Welch JL, Rossetti BJ, Rieken CW, Dewhirst FE, Borisy GG. Biogeography of a human oral microbiome at the micron scale. Proc Natl Acad Sci U S A. 2016;113(6):E791–800.
13. Kolenbrander PE, Palmer RJ Jr, Rickard AH, Jakubovics NS, Chalmers NI, Diaz PI. Bacterial interactions and successions during plaque development. Periodontology 2000. 2006;42:47–79.
14. Kolenbrander PE, Palmer RJ Jr, Periasamy S, Jakubovics NS. Oral multispecies biofilm development and the key role of cell-cell distance. Nat Rev Microbiol. 2010;8(7):471–80.
15. Kreth J, Herzberg MC. Molecular principles of adhesion and biofilm formation. In: de Paz LEC, Sedgley CM, Kishen A, editors. The root canal biofilm. Berlin: Springer; 2015. p. 23–54.
16. Tanaka H, Ebara S, Otsuka K, Hayashi K. Adsorption of saliva-coated and plain streptococcal cells to the

surfaces of hydroxyapatite beads. Arch Oral Biol. 1996;41(5):505–8.
17. Dawes C, Pedersen AM, Villa A, Ekstrom J, Proctor GB, Vissink A, Aframian D, McGowan R, Aliko A, Narayana N, et al. The functions of human saliva: a review sponsored by the World Workshop on Oral Medicine VI. Arch Oral Biol. 2015;60(6):863–74.
18. Nikitkova AE, Haase EM, Scannapieco FA. Taking the starch out of oral biofilm formation: molecular basis and functional significance of salivary alpha-amylase binding to oral streptococci. Appl Environ Microbiol. 2013;79(2):416–23.
19. Delius J, Trautmann S, Medard G, Kuster B, Hannig M, Hofmann T. Label-free quantitative proteome analysis of the surface-bound salivary pellicle. Colloids Surf B Biointerfaces. 2017;152:68–76.
20. Tabak LA. In defense of the oral cavity: structure, biosynthesis, and function of salivary mucins. Annu Rev Physiol. 1995;57:547–64.
21. Plummer C, Wu H, Kerrigan SW, Meade G, Cox D, Ian Douglas CW. A serine-rich glycoprotein of Streptococcus sanguis mediates adhesion to platelets via GPIb. Br J Haematol. 2005;129(1):101–9.
22. Slomiany BL, Murty VL, Piotrowski J, Slomiany A. Salivary mucins in oral mucosal defense. Gen Pharmacol. 1996;27(5):761–71.
23. Caufield PW, Dasanayake AP, Li Y, Pan Y, Hsu J, Hardin JM. Natural history of Streptococcus sanguinis in the oral cavity of infants: evidence for a discrete window of infectivity. Infect Immun. 2000;68(7):4018–23.
24. Frenkel ES, Ribbeck K. Salivary mucins in host defense and disease prevention. J Oral Microbiol. 2015;7:29759.
25. Gradinaru I, Ghiciuc CM, Popescu E, Nechifor C, Mandreci I, Nechifor M. Blood plasma and saliva levels of magnesium and other bivalent cations in patients with parotid gland tumors. Magnes Res. 2007;20(4):254–8.
26. Deng L, Bensing BA, Thamadilok S, Yu H, Lau K, Chen X, Ruhl S, Sullam PM, Varki A. Oral streptococci utilize a Siglec-like domain of serine-rich repeat adhesins to preferentially target platelet sialoglycans in human blood. PLoS Pathog. 2014;10(12):e1004540.
27. Bridier A, Piard JC, Pandin C, Labarthe S, Dubois-Brissonnet F, Briandet R. Spatial organization plasticity as an adaptive driver of surface microbial communities. Front Microbiol. 2017;8:1364.
28. Diaz PI, Chalmers NI, Rickard AH, Kong C, Milburn CL, Palmer RJ Jr, Kolenbrander PE. Molecular characterization of subject-specific oral microflora during initial colonization of enamel. Appl Environ Microbiol. 2006;72(4):2837–48.
29. Lemos JA, Burne RA. A model of efficiency: stress tolerance by Streptococcus mutans. Microbiology. 2008;154(Pt 11):3247–55.
30. Liu Y, Palmer SR, Chang H, Combs AN, Burne RA, Koo H. Differential oxidative stress tolerance of Streptococcus mutans isolates affects competition in an ecological mixed-species biofilm model. Environ Microbiol Rep. 2018;10(1):12–22.
31. Wilking JN, Zaburdaev V, De Volder M, Losick R, Brenner MP, Weitz DA. Liquid transport facilitated by channels in Bacillus subtilis biofilms. Proc Natl Acad Sci U S A. 2013;110(3):848–52.
32. Hwang G, Liu Y, Kim D, Sun V, Aviles-Reyes A, Kajfasz JK, Lemos JA, Koo H. Simultaneous spatiotemporal mapping of in situ pH and bacterial activity within an intact 3D microcolony structure. Sci Rep. 2016;6:32841.
33. Roberts AP, Kreth J. The impact of horizontal gene transfer on the adaptive ability of the human oral microbiome. Front Cell Infect Microbiol. 2014;4:124.
34. Fontaine L, Wahl A, Flechard M, Mignolet J, Hols P. Regulation of competence for natural transformation in streptococci. Infect Genet Evol. 2015;33:343–60.
35. Benitez-Paez A, Belda-Ferre P, Simon-Soro A, Mira A. Microbiota diversity and gene expression dynamics in human oral biofilms. BMC Genomics. 2014;15:311.
36. Edlund A, Garg N, Mohimani H, Gurevich A, He X, Shi W, Dorrestein PC, McLean JS. Metabolic fingerprints from the human oral microbiome reveal a vast knowledge gap of secreted small peptidic molecules. mSystems. 2017;2(4):e00058-17. https://doi.org/10.1128/mSystems.00058-17.
37. Koo H, Allan RN, Howlin RP, Stoodley P, Hall-Stoodley L. Targeting microbial biofilms: current and prospective therapeutic strategies. Nat Rev Microbiol. 2017;15(12):740–55.
38. Zhang JF, Wu R, Fan Y, Liao S, Wang Y, Wen ZT, Xu X. Antibacterial dental composites with chlorhexidine and mesoporous silica. J Dent Res. 2014;93(12):1283–9.
39. Buxadera-Palomero J, Canal C, Torrent-Camarero S, Garrido B, Javier Gil FJ, Rodríguez D. Antifouling coatings for dental implants: polyethylene glycol-like coatings on titanium by plasma polymerization. Biointerphases. 2015;10(2):029505.
40. Zhang Q, Nijampatnam B, Hua Z, Nguyen T, Zou J, Cai X, Michalek SM, Velu SE, Wu H. Structure-based discovery of small molecule inhibitors of cariogenic virulence. Sci Rep. 2017;7(1):5974.
41. Apel C, Barg A, Rheinberg A, Conrads G, Wagner-Döbler I. Dental composite materials containing carolacton inhibit biofilm growth of Streptococcus mutans. Dent Mater. 2013;29(11):1188–99.
42. Cheng L, Weir MD, Xu HHK, Antonucci JM, Kraigsley AM, Lin NJ, Lin-Gibson S, Zhou X. Antibacterial amorphous calcium phosphate nanocomposites with a quaternary ammonium dimethacrylate and silver nanoparticles. Dent Mater. 2012a;28(5):561–72.
43. Li F, Wang P, Weir MD, Fouad AF, Xu HHK. Evaluation of antibacterial and remineralizing nanocomposite and adhesive in rat tooth cavity model. Acta Biomater. 2014a;10(6):2804–13.

44. Li F, Weir MD, Xu HHK. Effects of quaternary ammonium chain length on antibacterial bonding agents. J Dent Res. 2013;92(10):932–8.
45. Cheng L, Zhang K, Zhang N, Melo MAS, Weir MD, Zhou XD, Bai YX, Reynolds MA, Xu HHK. Developing a new generation of antimicrobial and bioactive dental resins. J Dent Res. 2017;96(8):855–63.
46. Li F, Weir MD, Chen J, Xu HHK. Effect of charge density of bonding agent containing a new quaternary ammonium methacrylate on antibacterial and bonding properties. Dent Mater. 2014b;30(4):433–41.
47. Liang X, Söderling E, Liu F, He J, Lassila LVJ, Vallittu PK. Optimizing the concentration of quaternary ammonium dimethacrylate monomer in bis-GMA/TEGDMA dental resin system for antibacterial activity and mechanical properties. J Mater Sci Mater Med. 2014;25(5):1387–93.
48. Zhou H, Liu H, Weir MD, Reynolds MA, Zhang K, Xu HHK. Three-dimensional biofilm properties on dental bonding agent with varying quaternary ammonium charge densities. J Dent. 2016;53:73–81.
49. Makvandi P, Ghaemy M, Mohseni M. Synthesis and characterization of photo-curable bis-quaternary ammonium dimethacrylate with antimicrobial activity for dental restoration materials. Eur Polym J. 2016;74:81–90.
50. Wang S, Zhou C, Ren B, Li X, Weir MD, Masri RM, Oates TW, Cheng L, Xu HKH. Formation of persisters in Streptococcus mutans biofilms induced by antibacterial dental monomer. J Mater Sci Mater Med. 2017;28(11):178.
51. Zhang N, Ma J, Melo MAS, Weir MD, Bai Y, Xu HHK. Protein-repellent and antibacterial dental composite to inhibit biofilms and caries. J Dent. 2015a;43(2):225–34.
52. Zhang N, Weir MD, Romberg E, Bai Y, Xu HHK. Development of novel dental adhesive with double benefits of protein-repellent and antibacterial capabilities. Dent Mater. 2015b;31(7):845–54.
53. Cao B, Tang Q, Li L, Lee CJ, Wang H, Zhang Y, Castaneda H, Cheng G. Integrated zwitterionic conjugated poly(carboxybetaine thiophene) as a new biomaterial platform. Chem Sci. 2015;6(1):782–8.
54. Cao Z, Mi L, Mendiola J, Ella-Menye JR, Zhang L, Xue H, Jiang S. Reversibly switching the function of a surface between attacking and defending against bacteria. Angew Chem Int Ed. 2012;51(11):2602–5.
55. Khanal M, Raks V, Issa R, Chernyshenko V, Barras A, Garcia Fernandez JM, Mikhalovska LI, Turcheniuk V, Zaitsev V, Boukherroub R, et al. Selective antimicrobial and antibiofilm disrupting properties of functionalized diamond nanoparticles against Escherichia coli and Staphylococcus aureus. Part Part Syst Charact. 2015;32(8):822–30.
56. Turcheniuk V, Turcheniuk K, Bouckaert J, Barras A, Dumych T, Bilyy R, Zaitsev V, Siriwardena A, Wang Q, Boukherroub R, et al. Affinity of glycan-modified nanodiamonds towards lectins and uropathogenic Escherichia coli. ChemNanoMat. 2016;2(4):307–14.
57. Wehling J, Dringen R, Zare RN, Maas M, Rezwan K. Bactericidal activity of partially oxidized nanodiamonds. ACS Nano. 2014;8(6):6475–83.
58. Lewis JS, Gittard SD, Narayan RJ, Berry CJ, Brigmon RL, Ramamurti R, Singh RN. Assessment of microbial biofilm growth on nanocrystalline diamond in a continuous perfusion environment. J Manuf Sci E T ASME. 2010;132(3):0309191–7.
59. Kaushik SN, Scoffield J, Andukuri A, Alexander GC, Walker T, Kim S, Choi SC, Brott BC, Eleazer PD, Lee JY, et al. Evaluation of ciprofloxacin and metronidazole encapsulated biomimetic nanomatrix gel on Enterococcus faecalis and Treponema denticola. Biomater Res. 2015;19(1):9.
60. Nijampatnam B, Casals L, Zheng R, Wu H, Velu SE. Hydroxychalcone inhibitors of Streptococcus mutans glucosyl transferases and biofilms as potential anticaries agents. Bioorg Med Chem Lett. 2016;26(15):3508–13.
61. Zhang Q, Nguyen T, McMichael M, Velu SE, Zou J, Zhou X, Wu H. New small-molecule inhibitors of dihydrofolate reductase inhibit Streptococcus mutans. Int J Antimicrob Agents. 2015c;46(2):174–82.
62. Hwang G, Koltisko B, Jin X, Koo H. Nonleachable imidazolium-incorporated composite for disruption of bacterial clustering, exopolysaccharide-matrix assembly, and enhanced biofilm removal. ACS Appl Mater Interfaces. 2017;9(44):38270–80.
63. Guo L, McLean JS, Yang Y, Eckert R, Kaplan CW, Kyme P, Sheikh O, Varnum B, Lux R, Shi W, et al. Precision-guided antimicrobial peptide as a targeted modulator of human microbial ecology. Proc Natl Acad Sci U S A. 2015;112(24):7569–74.
64. Qian S, Cheng YF. Fabrication of micro/nanostructured superhydrophobic ZnO-alkylamine composite films on steel for high-performance self-cleaning and anti-adhesion of bacteria. Colloids Surf A Physicochem Eng Asp. 2018;544:35–43.
65. Michalska M, Gambacorta F, Divan R, Aranson IS, Sokolov A, Noirot P, Laible PD. Tuning antimicrobial properties of biomimetic nanopatterned surfaces. Nanoscale. 2018;10(14):6639–50.
66. Wu S, Zuber F, Maniura-Weber K, Brugger J, Ren Q. Nanostructured surface topographies have an effect on bactericidal activity. J Nanobiotechnol. 2018;16(1):20.
67. Zhu C, Zhang WW, Fang SY, Kong R, Zou G, Bao NR, Zhao JN, Shang XF. Antibiotic peptide-modified nanostructured titanium surface for enhancing bactericidal property. J Mater Sci. 2018;53(8):5891–908.
68. Amend S, Frankenberger R, Lücker S, Domann E, Krämer N. Secondary caries formation with a two-species biofilm artificial mouth. Dent Mater. 2018;34(5):786–96.
69. Aljerf L, Choukaife AE. Hydroxyapatite and fluorapatite behavior with pH change. Int Med J. 2017;24(5):407–10.

70. Andrucioli MCD, Faria G, Nelson-Filho P, Romano FL, Matsumoto MAN. Influence of resin-modified glass ionomer and topical fluoride on levels of Streptococcus mutans in saliva and biofilm adjacent to metallic brackets. J Appl Oral Sci. 2017;25(2):196–202.
71. Featherstone JD, Fontana M, Wolff M. Novel anti-caries and remineralization agents: future research needs. J Dent Res. 2018;97(2):125–7.
72. Mayanagi G, Igarashi K, Washio J, Domon-Tawaraya H, Takahashi N. Effect of fluoride-releasing restorative materials on bacteria-induced pH fall at the bacteria-material interface: an in vitro model study. J Dent. 2014;42(1):15–20.
73. Naoum S, Ellakwa A, Martin F, Swain M. Fluoride release, recharge and mechanical property stability of various fluoride-containing resin composites. Oper Dent. 2011;36(4):422–32.
74. Wang Y, Samoei GK, Lallier TE, Xu X. Synthesis and characterization of new antibacterial fluoride-releasing monomer and dental composite. ACS Macro Lett. 2013;2(1):59–62.
75. Cheng L, Weir MD, Xu HHK, Kraigsley AM, Lin NJ, Lin-Gibson S, Zhou X. Antibacterial and physical properties of calcium-phosphate and calcium-fluoride nanocomposites with chlorhexidine. Dent Mater. 2012b;28(5):573–83.
76. Cochrane NJ, Cai F, Huq NL, Burrow MF, Reynolds EC. Critical review in oral biology & medicine: new approaches to enhanced remineralization of tooth enamel. J Dent Res. 2010;89(11):1187–97.
77. Rodrigues MC, Chiari MDS, Alania Y, Natale LC, Arana-Chavez VE, Meier MM, Fadel VS, Vichi FM, Hewer TLR, Braga RR. Ion-releasing dental restorative composites containing functionalized brushite nanoparticles for improved mechanical strength. Dent Mater. 2018;34:746–55.
78. Karlinsey RL, Mackey AC. Solid-state preparation and dental application of an organically modified calcium phosphate. J Mater Sci. 2009;44(1):346–9.
79. Kim DA, Lee JH, Jun SK, Kim HW, Eltohamy M, Lee HH. Sol–gel-derived bioactive glass nanoparticle-incorporated glass ionomer cement with or without chitosan for enhanced mechanical and biomineralization properties. Dent Mater. 2017;33(7):805–17.
80. Nommeots-Nomm A, Labbaf S, Devlin A, Todd N, Geng H, Solanki AK, Tang HM, Perdika P, Pinna A, Ejeian F, et al. Highly degradable porous melt-derived bioactive glass foam scaffolds for bone regeneration. Acta Biomater. 2017;57:449–61.
81. Fernando D, Attik N, Pradelle-Plasse N, Jackson P, Grosgogeat B, Colon P. Bioactive glass for dentin remineralization: a systematic review. Mater Sci Eng C. 2017;76:1369–77.
82. Bakri MM, Hossain MZ, Razak FA, Saqina ZH, Misroni AA, Ab-Murat N, Kitagawa J, Saub RB. Dentinal tubules occluded by bioactive glass-containing toothpaste exhibit high resistance toward acidic soft drink challenge. Aust Dent J. 2017;62(2):186–91.
83. Khvostenko D, Hilton TJ, Ferracane JL, Mitchell JC, Kruzic JJ. Bioactive glass fillers reduce bacterial penetration into marginal gaps for composite restorations. Dent Mater. 2016;32(1):73–81.
84. Lee SY, Regnault WF, Antonucci JM, Skrtic D. Effect of particle size of an amorphous calcium phosphate filler on the mechanical strength and ion release of polymeric composites. J Biomed Mater Res B Appl Biomater. 2007;80(1):11–7.
85. Marovic D, Tarle Z, Hiller KA, Müller R, Rosentritt M, Skrtic D, Schmalz G. Reinforcement of experimental composite materials based on amorphous calcium phosphate with inert fillers. Dent Mater. 2014;30(9):1052–60.
86. O'Donnell JNR, Schumacher GE, Antonucci JM, Skrtic D. Structure-composition-property relationships in polymeric amorphous calcium phosphate-based dental composites. Materials. 2009;2(4):1929–54.
87. Chen C, Weir MD, Cheng L, Lin NJ, Lin-Gibson S, Chow LC, Zhou X, Xu HHK. Antibacterial activity and ion release of bonding agent containing amorphous calcium phosphate nanoparticles. Dent Mater. 2014;30(8):891–901.
88. Chen Z, Cao S, Wang H, Li Y, Kishen A, Deng X, Yang X, Wang Y, Cong C, Wang H, et al. Biomimetic remineralization of demineralized dentine using scaffold of CMC/ACP nanocomplexes in an in vitro tooth model of deep caries. PLoS One. 2015;10(1):e0116553.
89. Cheng L, Zhang K, Zhou CC, Weir MD, Zhou XD, Xu HHK. One-year water-ageing of calcium phosphate composite containing nano-silver and quaternary ammonium to inhibit biofilms. Int J Oral Sci. 2016;8(3):172–81.
90. Zhang K, Cheng L, Weir MD, Bai YX, Xu HHK. Effects of quaternary ammonium chain length on the antibacterial and remineralizing effects of a calcium phosphate nanocomposite. Int J Oral Sci. 2016;8:45–53.
91. Laurent P, Camps J, About I. BiodentineTM induces TGF-β1 release from human pulp cells and early dental pulp mineralization. Int Endod J. 2012;45(5):439–48.
92. Hegde S, Sowmya B, Mathew S, Bhandi SH, Nagaraja S, Dinesh K. Clinical evaluation of mineral trioxide aggregate and biodentine as direct pulp capping agents in carious teeth. J Conserv Dent. 2017;20(2):91–5.
93. Simila HO, Karpukhina N, Hill RG. Bioactivity and fluoride release of strontium and fluoride modified biodentine. Dent Mater. 2018;34(1):e1–7.
94. Vural UK, Kiremitci A, Gokalp S. Randomized clinical trial to evaluate MTA indirect pulp capping in deep caries lesions after 24-months. Oper Dent. 2017;42(5):470–7.
95. Song M, Kang M, Kim HC, Kim E. A randomized controlled study of the use of proroot mineral trioxide aggregate and endocem as direct pulp capping materials. J Endod. 2015;41(1):11–5.

96. Harris D, Ummadi JG, Thurber AR, Allau Y, Verba C, Colwell F, Torres ME, Koley D. Real-time monitoring of calcification process by Sporosarcina pasteurii biofilm. The Analyst. 2016;141(10):2887–95.
97. Joshi VS, Kreth J, Koley D. Pt-decorated MWCNTs-ionic liquid composite-based hydrogen peroxide sensor to study microbial metabolism using scanning electrochemical microscopy. Anal Chem. 2017a;89(14):7709–18.
98. Joshi VS, Sheet PS, Cullin N, Kreth J, Koley D. Real-time metabolic interactions between two bacterial species using a carbon-based pH microsensor as a scanning electrochemical microscopy probe. Anal Chem. 2017b;89(20):11044–52.
99. Liu X, Ramsey MM, Chen X, Koley D, Whiteley M, Bard AJ. Real-time mapping of a hydrogen peroxide concentration profile across a polymicrobial bacterial biofilm using scanning electrochemical microscopy. Proc Natl Acad Sci U S A. 2011;108(7):2668–73.
100. Ummadi JG, Joshi VS, Gupta PR, Indra AK, Koley D. Single-cell migration as studied by scanning electrochemical microscopy. Anal Methods. 2015;7(20):8826–31.
101. Ummadi JG, Downs CJ, Joshi VS, Ferracane JL, Koley D. Carbon-based solid-state calcium ion-selective microelectrode and scanning electrochemical microscopy: a quantitative study of pH-dependent release of calcium ions from bioactive glass. Anal Chem. 2016;88(6):3218–26.
102. Stoica L, Neugebauer S, Schuhmann W. Scanning electrochemical microscopy (SECM) as a tool in biosensor research. Adv Biochem Eng Biotechnol. 2008;109:455–92.
103. Amemiya S, Bard AJ, Fan FR, Mirkin MV, Unwin PR. Scanning electrochemical microscopy. Annu Rev Anal Chem (Palo Alto, CA). 2008;1:95–131.
104. Roberts WS, Lonsdale DJ, Griffiths J, Higson SP. Advances in the application of scanning electrochemical microscopy to bioanalytical systems. Biosens Bioelectron. 2007;23(3):301–18.
105. Bard AJ, Mirkin MV. Scanning electrochemical microscopy. 2nd ed. Boca Raton, FL: CRC Press; 2012.

Multifunctional Restorative Dental Materials: Remineralization and Antibacterial Effect

9

Roberto Ruggiero Braga

Abstract

Remineralizing biomaterials have been a longtime pursuit in dentistry as a strategy to prevent or at least postpone the development of caries lesions around existing restorations, fissure sealants, and orthodontic brackets. Glass-ionomer cements, with a track record spanning four decades, have shown good results *in situ*. However, their low mechanical properties and bond strength to the tooth structure are limiting factors in several clinical situations. In the last decade, calcium orthophosphates (e.g., amorphous calcium phosphate/ACP), bioactive glasses (e.g., 45S5), and calcium silicates (e.g., mineral trioxide aggregate/MTA) have been tested as ion-releasing fillers in dentin bonding systems and resin composites. *In vitro* testing showed unequivocal evidences of hybrid layer remineralization, which reduces permeability and collagen degradation, therefore contributing to the longevity of bonded interfaces. On enamel, composites containing calcium orthophosphates were shown to promote mineral recovery *in vitro* and reduce mineral loss *in situ*. Besides fostering remineralization, some of these particles may also grant antimicrobial activity to resin-based materials, making them "multifunctional restorative materials." Studies show that bioactive glasses are effective against some bacterial species due to their alkalinity and effect on osmotic gradient. For calcium silicates, however, there seems to be no consensus among authors regarding antimicrobial effect, while calcium orthophosphates and glass-ionomers show no evidence of intrinsic antimicrobial activity.

R. R. Braga (✉)
Department of Biomaterials and Oral Biology, University of São Paulo School of Dentistry, São Paulo, Brazil
e-mail: rrbraga@usp.br

9.1 Introduction

Restorative materials with remineralizing and antibacterial properties are not unknown to dentistry. For instance, calcium hydroxide and glass-ionomer cements (GIC) have a very long history of clinical use. The last few decades, however, have seen a surge in research of new multifunctional resin-based materials combining both effects in a vast range of applications including desensitization of exposed cervical dentin, biomimetic dentin remineralization, atraumatic restorative treatment (ART), orthodontic cements, pulp capping, and as direct restorative materials. These new materials contain bioactive glass, calcium silicate, or calcium orthophosphate particles dispersed in a dimethacrylate-based resin matrix. Because the antibacterial effect is not necessarily the primary feature of these ion-releasing fillers, other antibacterial agents (e.g., chlorhexidine or silver nanoparticles) can be associated.

A. C. Ionescu, S. Hahnel (eds.), *Oral Biofilms and Modern Dental Materials*,
https://doi.org/10.1007/978-3-030-67388-8_9

Multifunctional materials are intended to promote apatite deposition in hard dental tissues (e.g., ART and remineralization of resin-infiltrated dentin) or to prevent the development of caries lesions by increasing mineral uptake after a demineralization event (e.g., bracket bonding and direct restorative materials). In either case, antibacterial activity is important to facilitate the intended outcome. Depending on the clinical situation, these effects are necessary only for relatively short periods of time or may be needed for as long as possible.

In this chapter, the current and most relevant findings on the research leading to the development of remineralizing and antibacterial materials are presented.

9.2 Calcium Orthophosphates

The first attempts of using calcium orthophosphates (CaP) as fillers in restorative composites date back to the 1980s. Interestingly, these studies focused not on remineralization, but on the development of coupling agents to improve adhesion to the tooth structure [1]. Later on, hydroxyapatite (HAP) particles were tested as reinforcing fillers, as its relatively low hardness could reduce composite wear damage [2]. It was also in the mid-1990s that amorphous calcium phosphate (ACP) started to be tested as bioactive filler in resin-based materials [3]. ACP is an intermediate phase in HAP precipitation and that, along with its relative solubility, makes it suitable as ion-releasing filler. Other orthophosphate phases, such as dicalcium phosphate anhydrous (DCPA), dicalcium phosphate dihydrate (DCPD), and tetracalcium phosphate (TTCP) are also found in the literature.

The development of remineralizing composites containing CaP particles is not without its drawbacks. For instance, the incorporation of CaP particles in dimethacrylate matrices leads to significant reductions in fracture strength due to the lack of a strong particle-resin interaction [4]. Therefore, there is a trade-off between bioactivity and mechanical strength [5]. Other mechanical properties, such as elastic modulus and fracture toughness, are less sensitive to the presence of CaP particles [6]. In order to improve the interaction between CaP particles and resin matrix and minimize the loss in strength, it is possible to functionalize these particles with organic molecules, such as carboxylic acids, silanes, or dimethacrylates [7–9]. These molecules work as coupling agents, binding to calcium and copolymerizing with the monomers in the matrix, and also improving the wettability of the resin on the particles.

Another point of concern is the long-term degradation of these materials. Since calcium and phosphate release occurs at the expense of particle surface dissolution, it is licit to assume that over time oral fluids would find opened pathways at the filler-matrix interfaces to penetrate the material and increase matrix degradation. There are very few studies that investigated this topic, with contradictory findings. While a more severe degradation was verified in composites containing DCPD after 28 days in water in relation to the control material [6], no differences were observed due to the presence of ACP after 2 years in water [10].

9.2.1 Remineralization Studies

Resin-based materials foster remineralization by releasing calcium and phosphate ions in supersaturating levels. Also, the presence of calcium in the biofilm increases fluoride retention, which also helps to prevent demineralization [11]. Ion release is determined by a number of factors, such as solubility of the calcium orthophosphate phase [12], particle surface area [9], CaP volume fraction in the composite [13], hydrophilicity of the resin matrix [14], and pH of the immersion medium [15]. *In vitro* studies have demonstrated that ionic concentrations released by CaP-containing composites are capable of promoting apatite precipitation [16, 17]. However, ion release does not occur indefinitely. Experimental composites containing 20 wt% of ACP showed ion release up to 70 days under very acidic conditions (pH 4), which increase particle erosion and, consequently, boost ion release [18].

Composites containing calcium orthophosphate particles were shown to promote mineral recovery in enamel artificial caries lesions *in vitro*. Small fractions of hydrophilic monomers such as 2-hydroxyethyl methacrylate (HEMA) and methacryloyloxyethyl phthalate (MEP) were added to the matrix to enhance fluid access to the particles. A dimethacrylate-based material containing 40 wt% of ACP (without reinforcing fillers) was able to recover 38% of the mineral content of the lesion (quantified by transverse microradiography, TMR) and reduce lesion depth by 23% after 2 weeks of pH cycling. In the same study, a composite containing HAP was as ineffective as the control composite with silica particles due to its reduced solubility [19, 20]. When compared to a commercial orthodontic cement containing fluoride, an ACP composite (40 wt%) promoted a mineral recovery of 14%, against 4% of the fluoride composite. Interestingly, in the top third of the lesion the mineral gain was higher for the fluoride material, while the ACP composite promoted higher mineral deposition at the deeper regions [19, 21]. Another study showed that after 30 days of pH cycling, an experimental composite containing 40 wt% of ACP and 20 wt% of reinforcing glass promoted 22% of mineral gain, in comparison to 6% of a commercial restorative composite containing ytterbium trifluoride [22].

The experimental model utilized in the studies mentioned in the previous paragraph does not truly represent the intended use of these materials, though it does serve as a "proof of concept" [21]. In a more clinically relevant *in situ* experiment, the protective effect of an ACP-containing composite on the surrounding enamel was verified as mineral loss and lesion depth was significantly lower than around a conventional composite after 14 days. Also, calcium and phosphate concentrations in the biofilm formed on the specimens were statistically higher [23].

In a series of studies, a two-paste resin cement containing approximately 40 wt% of tetracalcium phosphate (TTCP) and dicalcium phosphate anhydrous (DCPA) intended for indirect pulp capping or atraumatic restorative treatment (ART) material was tested *in vitro* and *in vivo*. Under static conditions (immersion in saliva-like solution, SLS, for 5 weeks), the cement was shown to promote a 38% recovery in dentin mineral content [24]. The use of a bonding agent between the demineralized dentin and the cement reduced remineralization, possibly due to calcium binding by the acidic monomers in the adhesive [25]. *In vivo*, the cement was applied directly on caries-affected dentin and under a conventional resin composite. After 3 months, calcium and phosphorous content was significantly higher in the treated dentin in comparison to the untreated control and similar to sound dentin levels up to a 30 μm depth [26].

Finally, in another example of *in vitro* top-down dentin remineralization, experimental composites containing 40 wt% of ACP and 20 wt% of either silanated glass or TTCP promoted 43–48% mineral recovery after 8 weeks of pH cycling [27].

9.2.2 Antibacterial Activity

The addition of calcium orthophosphate particles in resin-based materials does not seem to provide any significant protection against biofilm formation. For example, the aforementioned *in situ* study found no reduction in the number of *Streptococci* and *Lactobacilli* colony-forming units (CFUs) grown on an ACP-containing composite in relation to the control [23]. Notwithstanding, the same material showed some acid-neutralizing activity, promoting a raise in pH from 4 to 7, which could reduce the growth of acidogenic bacteria [28]. The buffering capacity of CaP-containing resin materials, though insufficient to reduce biofilm growth, was confirmed in a subsequent study [29].

Multifunctional composites and bonding agents associating calcium orthophosphates with quaternary ammonium monomers [30–32], silver (Ag) nanoparticles [32, 33], or chlorhexidine [34] were tested with overall good results in terms of antibacterial activity. Particularly in the case of adhesive systems, antibacterial agents can be added to both the primer and the bonding resin, while ACP was added to the latter. Bond strength tests showed that the addition of antibacterial agents and calcium phosphate particles did not reduce bond strength after 28-day storage [31]. Transmission electron microscopy revealed the presence of ACP agglomerates and Ag nanoparticles in the resin tags [35].

Recently, silver phosphate/calcium phosphate particles were synthesized. These particles are capable of producing metallic Ag nanoparticles *in situ* when exposed to UV-Vis radiation (<530 nm), therefore, in the range emitted by dental light-curing units. Calcium release from resin materials containing 20–30 wt% of these mixed phosphate particles was similar to that of calcium phosphate only, while *S. mutans* CFU count was reduced by three log units [36].

9.3 Bioactive Glasses

Silica-based (SiO_2) glasses have been widely studied in the past 50 years, after Hench and colleagues found out that certain compositions can chemically bond to bone [37]. The most studied glass composition (Bioglass™ 45S5, $45SiO_2$, 24.5CaO, $24.5Na_2O$, $6P_2O_5$, in wt%; "5" meaning a 5:1 calcium-to-phosphorus molar ratio) showed good results when used for pulp capping in animal models [38] and is found in products indicated for the repair of alveolar bone defects [39] and treatment of dentin hypersensitivity [40].

Similarly, 45S5 glass plates were shown to bond to etched dentin (35% phosphoric acid for 15 s) after 3 weeks in artificial saliva [41]. The basic bonding mechanism can be summarized as follows: at initial stages of glass dissolution, side groups on type I collagen fibers can bind to the negatively charged particle surface; at later stages, an interfacial layer of hydroxycarbonate apatite (HAC) nucleates on top of a silica gel layer containing silanol groups (Si–OH), with interpenetrating collagen fibers [42]. Interestingly, apatite-wollastonite (A/W) glass-ceramic (4.6MgO, 44.9CaO, $16.3P_2O_5$, $34.2SiO_2$, in wt%) did not bond to dentin. Other bioactive glass (BAG) compositions, such as S53P4 ($53SiO_2$, $23Na_2O$, 20CaO, and $4P_2O_5$, in wt%), were capable of promoting HAC deposition and obliterate the dentin tubules after 24-h immersion in a BAG suspension followed by 2 weeks of incubation [43].

9.3.1 Remineralization Studies

BAGs have been tested for dentin remineralization with promising results. Demineralized dentin bars were shown to recover the carbon-to-mineral ratio (determined by thermal analyses) of natural dentin after 30-day immersion in a suspension of nanometric 45S5-type particles (30–50 nm, surface area: 64 m^2/g). However, the flexural strength and elastic modulus remained similar to those of demineralized dentin [44]. In fact, more recent studies have demonstrated that acidic polymers are necessary as biomimetic precursors in order to guide dentin remineralization and recover its mechanical properties (i.e., bottom-up remineralization) [45]. Apatite deposition, with obliteration of dentinal tubules, was also observed on demineralized dentin samples treated with 45S5 (20 mg for 1 min) after 7-day storage in artificial saliva [46].

Adhesive systems containing BAGs have been tested as a way to reduce the long-term degradation of the bonded interface. The incorporation of 30 wt% of 45S5 particles (<10 μm) in a Bis-GMA/HEMA resin adhesive was able to maintain the microtensile bond strength to dentin after 6 months in phosphate-buffered solution (PBS), compared to a 37% reduction displayed by the control adhesive [47]. One of the mechanisms proposed to explain the lower degradation was that mineral deposition would replace water-rich domains within the hybrid layer, reducing hydrolysis. The precipitation of calcium phosphates also interferes with metalloproteinase (MMP) and cathepsin activity, reducing enzymatic degradation of the collagen [48, 49].

Apatite deposition from composites containing both silanized reinforcing glass and BAG particles was also tested. When 15 wt% of BAG particles with Ca/P = 4 was added to a commercial flowable composite, apatite formation was verified after 20 days in simulated body fluid (SBF) [50]. Experimental resins containing 37.5–50.0 wt% of 45S5 were shown to prevent enamel demineralization after 45-day immersion in lactic acid. The proposed mechanism was acid neutralization by the ions released from the glass [51].

Similarly to what was described for CaP-containing composites, the effect of bioactive glass particles on the mechanical properties of experimental composites is a clinically relevant concern. The replacement of reinforcing fillers by 10 or 15 wt% of bioactive glass (65% SiO_2, 31% CaO, 4% P_2O_5 in mols, particle size: 0.04–3.0 μm) did not lead to significant reductions in flexural strength or fracture toughness in comparison to the control composite. Also, the presence of bioactive glass did not increase composite degradation after 2-month immersion in brain-heart infusion medium [52].

9.3.2 Antibacterial Effect

The antibacterial effect of bioactive glasses is a topic of great interest in orthopedics, as a way to prevent medical device-associated infections (MDAIs) in joint and bone implant surgeries [53], and a consensus seems to exist among authors about BAG efficacy against bacteria. Bioactive glasses are considered materials with intrinsic antimicrobial activity due to the release of ions such as Na^+ and Ca^{2+} that leads to a local increase in osmotic concentration and pH. As a result, there is an unbalance in bacterial intracellular Ca^{2+}, leading to membrane depolarization and bacterial death [54]. In dentistry, S53P4 (20 μm) was tested against cultures of Gram-positive (*A. naeslundii*, *S. mutans*, and *S. sanguis*) and Gram-negative pathogens (*A. actinomycetemcomitans* and *P. gingivalis*). Except for *S. sanguis*, the other species lost viability after 60 min of incubation in the presence of S53P4. Besides a pH raise to 10.8, the increase in osmotic pressure resulting from Na^+ release and bacterial agglutination (for *P. gingivalis*) in the presence of Ca^{2+} are also listed as antibacterial mechanisms [55]. A similar increase in pH was verified for 45S5 glass (90–710 μm), which led to a 93–99% reduction in supragingival bacterial cultures after 3 h of incubation (*A. viscous*, *S. mutans*, and *S. sanguis*). Subgingival bacterial species (*A. actinomycetemcomitans*, *F. nucleatum*, *Prev. intermedia*, and *P. gingivalis*) presented 91–100% of reduction in viability, while the control (non-bioactive) glass resulted in reductions of 8–62% [56].

An experimental composite containing 15 wt% of bioactive glass (65SiO_2, 31CaO, and 4P_2O_5, in mols, particle size: 0.04–3 μm) was able to reduce bacterial penetration along the tooth/restoration interface. While for the control composite the entire axial wall was infiltrated by *S. mutans* after 2 weeks in a bioreactor under cyclic loading, for the composite with bioactive glass, only 61% of the gap depth showed bacterial penetration [57].

Bioactive glasses can have their antibacterial effect enhanced by the incorporation of metallic elements such as silver, copper, strontium, or zinc. Among them, Ag-doped glasses are the most studied [53]. Ag-doped bioactive glass particles (58.6SiO_2, 24.9CaO, 7.2P_2O_5, 4.2Al_2O_3, 1.5Na_2O, 1.5K_2O, 2.1Ag_2O, in wt%, particle size: 25 μm) were incorporated into a commercial flowable composite. After 8 days in PBS, the extract of the composite containing 15 wt% of Ag-doped glass was able to completely inhibit *S. mutans* growth. Interestingly, PBS pH (7.4) was not affected by the immersion of the composite specimens, indicating that antibacterial activity was due to Ag release from the material [50].

9.4 Calcium Silicates

Calcium silicates share some of the characteristics and mechanisms described for bioactive glasses, in terms of both remineralization and antimicrobial activity. The use of calcium silicates in dentistry gained mommentum in the mid-1990s, with the development of MTA cements (mineral trioxide aggregate) for use in endodontics as root-end and furcal perforation filling material [58]. MTA is a mixture of dicalcium silicate (belite), tricalcium silicate (alite), tricalcium aluminate and tetracalcium aluminoferrite (Portland cement, amounting to 75% of the entire mass), bismuth oxide (20%), calcium sulfate (gypsum, 5%), and trace amounts of other metallic silicates and oxides. Hydration of MTA produces a calcium silicate hydrate gel and crystalline calcium hydroxide [59].

The good results shown by calcium silicates in several applications, including as pulp capping materials, may explain why much of the research on dentin remineralization produced in the last few years has focused on these particles as ion source. Unfortunately, no reports have been published on the mechanical properties of resin-based materials containing calcium silicates or reinforcing glass fillers associated with calcium silicates. Therefore, their reinforcing effect on the resin matrix is still to be verified. Experimental materials containing no reinforcing fillers and 56 wt% of calcium aluminosilicate particles in a light-curable hydrophilic resin matrix showed water sorption four times higher than a commercial flowable composite, and similar to a resin-modified glass-ionomer [60].

9.4.1 Remineralization Studies

The availability of calcium ions in a highly alkaline environment created by the hydroxyl ions, associated with the phosphate present in physiologic fluids, favors the precipitation of poorly crystalline calcium-deficient carbonated apatite [61]. This characteristic was first identified in relation to calcium silicates' sealing ability when used as root-end filling material [62]. When applied to mineral-depleted dentin, calcium silicates associated with analogues of acidic non-collagenous proteins (e.g., polyacrylic acid) allow for the production of metastable amorphous calcium phosphate nanoprecursors and, at a later stage, intrafibrillar and interfibrillar apatite deposition (biomimetic, "bottom-up" remineralization) [63]. Evidences of remineralization within the hybrid layer kept in contact with Portland cement discs were detected after 2–4 months of immersion in SBF containing polyacrylic acid and polyvinylphosphonic acid [64].

The incorporation of Portland cement-based particles in experimental adhesives was able to maintain the bond strength of dentin-composite interfaces after 6 months in SBS. According to the authors, besides mineral deposition, the increase in pH within the hybrid layer may interfere with MMP activity [65]. The increase in nanohardness and elastic modulus of the hybrid layer and the reduction in micropermeability as evidences of remineralization were also observed after 3 months in SBS [66]. It is important to remember that, since calcium silicates do not have phosphorus in their composition as they rely on external phosphate sources to promote apatite deposition [48].

The efficacy of calcium silicates in relation to bioactive glasses was evaluated in several studies. For example, bioactive glass 45S5 was found to inhibit MMP activity in dentin demineralized by both phosphoric acid and EDTA, while particles containing 90% of Portland cement and 10% of β-tricalcium phosphate were only efficient when applied to EDTA-treated dentin. Since EDTA is not capable of removing all the phosphoproteins from the collagen, the residual phosphate favored calcium phosphate precipitation in samples infiltrated with modified calcium silicate particles. Calcium phosphates, in turn, are capable of inhibiting MMP activity, as well as forming CaP:MMP complexes with restricted mobility, preventing collagen enzymatic degradation [67]. In another study, discs of resin containing 33 wt% of polycarboxylated bioactive glass or Portland cement were kept in contact with dentin samples, immersed in artificial saliva. After 14 days, no difference in remineralization was observed between both groups [68]. The use of experimental adhesives containing bioactive glass or MTA particles (40 wt%) resulted in similar bond strength values after 10 months of storage in phosphate-buffered saline, both statistically higher than the control adhesive without ion-releasing particles, ascribed to the maintenance of the hybrid-layer integrity [69].

9.4.2 Antimicrobial Effect

Most of the available literature evaluating the antimicrobial effect of calcium silicates tested MTA against species usually identified in periapical and root infections. Similar to bioactive glasses, MTA also shows a rapid increase in pH after mixing due to the formation of calcium hydroxide as one of the hydration products. Its initial pH is 10.2, reaching a plateau at 12.5 after 3 h [58].

In spite of its alkalinity, MTA antimicrobial activity is controversial. For example, an agar diffusion test showed no inhibitory effect over seven anaerobic bacteria and limited effect on five out of nine facultative bacteria usually found in infected root canals, either immediately or 24 h after mixing [70]. In another study using the direct contact test, MTA showed growth inhibition of *E. faecalis* and *C. albicans* when placed in the culture media 20 min or 1 day after mixing [71]. The reason for the lack of consensus in the literature has been attributed to differences in methods and microbial strains as well as material-related variables, such as power-to-liquid ratio and source of the MTA [72, 73]. An interesting finding was reported where authors found that *E. faecalis* inhibition increased when specimens made of crushed set MTA were incubated with dentin powder [74]. This phenomenon had been reported in relation to S53P4 glass and seems to be related to a higher dissolution rate of the particles in the presence of dentin powder [75].

In order to improve their antimicrobial activity, calcium silicates have been associated with other compounds. Ag-doped and chlorhexidine-loaded calcium silicate nanoparticles showed good substantivity against *E. faecalis* due to its retention on the dentin surface by means of an apatite layer between the nanoparticles and the dentin surface [76, 77]. A quaternary ammonium monomer (QAM), 2-methacryloxylethyl dodecyl methyl ammonium bromide (MAE-DB), added to an experimental resin containing Portland cement particles showed significant antibacterial activity against *S. mutans* [78].

9.5 Glass-Ionomers

Glass-ionomer cements (GIC) are acid-base cements with widespread use in dentistry as restorative and luting material, orthodontic cement, and sealant. Their mechanical properties, however, are not high enough to allow their use in large cavities on stress-bearing areas. Their remineralizing and antibacterial effects are attributed to the presence of fluoride in the silicate-based glass particles, which is initially released upon their reaction with the polyalkenoic acids in the cement liquid and, in smaller concentrations, over time due to particle dissolution. From the cariology standpoint, it is important to point out that fluoride does not prevent caries lesion development, but it does slow down its progression. In fact, the incorporation of fluoride into enamel and dentin is a consequence of the caries process [79]. Still, the possibility of remineralizing caries-affected dentin, associated with ease of use and good marginal sealing, makes GIC the material of choice in atraumatic restorative treatment (ART) techniques.

There is a vast amount of literature on remineralization and antibacterial properties of GIC cements evaluated *in vitro* and *in situ* but, unfortunately, the clinical evidences are scant and most often point to the absence of significant effects of fluoride-releasing materials regarding the prevention of caries lesion development.

9.5.1 Remineralization Studies

GICs, both conventional and resin-modified versions, were shown to reduce enamel demineralization *in situ*. When pH drops below 5.5 due to acid production by bacteria in the biofilm, HAP dissolution takes place. In the presence of fluoride, however, this process is counteracted by the deposition of fluorapatite, which does not dissociate at pH values above 4.5. Therefore, increases in fluoride content in the enamel adjacent to the restoration are actually the result of the de- and remineralization process, and should not be regarded as an indication of an enamel-"strengthening effect" granted by the fluoride-releasing material [79]. An important aspect that must be taken into consideration in *in situ* studies evaluating GIC is the association or not with other sources of fluoride. When GIC restorations were evaluated in patients making use of fluoridated dentifrices, no differences in plaque fluoride or mineral loss around the restoration were observed in comparison to a resin composite [80, 81].

Nevertheless, fluoride release from GIC seems to be an effective way to increase mineral content of caries-affected dentin in ART procedures. *In vivo* studies revealed that fluoride from GIC penetrates partially demineralized, caries-affected dentin through an ion-exchange process taking place to buffer the low pH of the fresh cement. This process seems to be driven by a concentration gradient between the GIC and the demineralized dentin, where fluoride and strontium (if present in the cement particles) would precipitate within the demineralized dentin [82, 83]. However, ultrastructural studies failed to encounter evidences of actual remineralization rather than simple mineral uptake [84].

9.5.2 Antibacterial Effect

Studies evaluating the effect of GIC on biofilm formation may show contradictory results due to differences in test methods and, most importantly, to aging conditions of the specimens. In general, freshly mixed cements show antibacterial effect due to its initial high fluoride release and low pH [85, 86]. However, this effect is lost with time. For instance, biofilm collected from the surface of aged (1 year) resin-modified glass-ionomer, compomer, resin composite, and intact enamel *in vivo* showed similar counts for streptococci and lactobacilli [87].

Fluoride 0.53 mmol/L (10 ppm) was shown to change biofilm composition and reduce *S. mutans* count in the presence of glucose due to a direct inhibition of its metabolism, which reduces acid production and favors the growth of less aciduric species [88]. However, it is unlikely that a material could provide such levels of fluoride to the biofilm in the long term [79], as fluoride concentration in the biofilm formed *in situ* on resin-modified GIC or on dentin after acidulated phosphate fluoride application showed values not higher than 0.01 mmol/L [81, 89].

The antibacterial effect of fluoride released from GIC is considered secondary to its effect on demineralization and the clinical effectiveness of high fluoride levels on the biofilm metabolism is unclear at best [90]. In order to increase the antibacterial effect of GIC, several approaches have been tested, including well-known antibacterial agents such as chlorhexidine [91], antibiotics [92], titanium oxide nanoparticles [93], chitosan [94], and silver nanoparticles [95].

9.6 Final Remarks

There are a multitude of compositional variables involved in the performance of remineralizing/antibacterial restorative materials. Among the ion-releasing particles being investigated, bioactive glasses (45S5 in particular) and calcium silicates show good results regarding dentin remineralization. On the other hand, most of the research on calcium orthophosphates focuses on their use in enamel remineralization. Overall, the bioactivity of these materials is expected to decrease over time, as observed with glass-ionomers, which may limit their clinical effectivity in some applications.

These ion-releasing fillers show different levels of antibacterial activity, granted by their effect on the osmotic gradient and alkalinity. Among them, bioactive glasses seem to be the most effective, while calcium orthophosphates seem to have a very limited antibacterial effect. Several antibacterial agents can be associated with these particles to enhance antibacterial activity.

The intense research on multifunctional restorative materials reveals their potential in several clinical applications. Notwithstanding, it is of fundamental importance to explore the long-term performance of these materials. The experience obtained with *in situ* and *in vivo* research on glass-ionomers emphasizes the importance of increasing the level of evidence in the near future, as evaluations conducted *in vitro* usually do not reproduce the complexity of the oral environment.

References

1. Antonucci JM, Misra DN, Peckoo RJ. The accelerative and adhesive bonding capabilities of surface-active accelerators. J Dent Res. 1981;60(7):1332–42.
2. Labella R, Braden M, Deb S. Novel hydroxyapatite-based dental composites. Biomaterials. 1994;15(15):1197–200.
3. Antonucci JM, Skrtic D, Eanes ED. Bioactive polymeric dental materials based on amorphous calcium phosphate. Polym Prepr. 1994;35(2):460–1.
4. Xu HH, Moreau JL. Dental glass-reinforced composite for caries inhibition: calcium phosphate ion release and mechanical properties. J Biomed Mater Res B Appl Biomater. 2010;92(2):332–40.
5. Alania Y, Chiari MD, Rodrigues MC, Arana-Chavez VE, Bressiani AH, Vichi FM, et al. Bioactive composites containing TEGDMA-functionalized calcium phosphate particles: degree of conversion, fracture strength and ion release evaluation. Dent Mater. 2016;32(12):e374–81.
6. Chiari MD, Rodrigues MC, Xavier TA, de Souza EM, Arana-Chavez VE, Braga RR. Mechanical properties and ion release from bioactive restorative composites containing glass fillers and calcium phosphate nano-structured particles. Dent Mater. 2015;31(6): 726–33.
7. Arcis RW, Lopez-Macipe A, Toledano M, Osorio E, Rodriguez-Clemente R, Murtra J, et al. Mechanical properties of visible light-cured resins reinforced with hydroxyapatite for dental restoration. Dent Mater. 2002;18(1):49–57.
8. Rodrigues MC, Hewer TL, Brito GE, Arana-Chavez VE, Braga RR. Calcium phosphate nanoparticles functionalized with a dimethacrylate monomer. Mater Sci Eng C Mater Biol Appl. 2014;45:122–6.
9. Xu HH, Weir MD, Sun L. Nanocomposites with Ca and PO4 release: effects of reinforcement, dicalcium phosphate particle size and silanization. Dent Mater. 2007;23(12):1482–91.
10. Moreau JL, Weir MD, Giuseppetti AA, Chow LC, Antonucci JM, Xu HH. Long-term mechanical durability of dental nanocomposites containing amorphous calcium phosphate nanoparticles. J Biomed Mater Res B Appl Biomater. 2012;100(5):1264–73.
11. Souza JG, Tenuta LM, Del Bel Cury AA, Nobrega DF, Budin RR, de Queiroz MX, et al. Calcium prerinse before fluoride rinse reduces enamel demineralization: an in situ caries study. Caries Res. 2016;50(4):372–7.
12. Rodrigues MC, Natale LC, Arana-Chaves VE, Braga RR. Calcium and phosphate release from resin-based materials containing different calcium orthophosphate nanoparticles. J Biomed Mater Res B Appl Biomater. 2015;103(8):1670–8.
13. Xu HH, Weir MD, Sun L, Takagi S, Chow LC. Effects of calcium phosphate nanoparticles on Ca-PO4 composite. J Dent Res. 2007;86(4):378–83.
14. Skrtic D, Antonucci JM. Dental composites based on amorphous calcium phosphate–resin composition/physicochemical properties study. J Biomater Appl. 2007;21(4):375–93.
15. Xu HH, Weir MD, Sun L. Calcium and phosphate ion releasing composite: effect of pH on release and mechanical properties. Dent Mater. 2009;25(4):535–42.
16. Aljabo A, Abou Neel EA, Knowles JC, Young AM. Development of dental composites with reactive fillers that promote precipitation of antibacterial-hydroxyapatite layers. Mater Sci Eng C Mater Biol Appl. 2016;60:285–92.
17. Skrtic D, Antonucci JM, Eanes ED. Improved properties of amorphous calcium phosphate fillers in remineralizing resin composites. Dent Mater. 1996;12(5):295–301.
18. Zhang L, Weir MD, Chow LC, Antonucci JM, Chen J, Xu HH. Novel rechargeable calcium phosphate dental nanocomposite. Dent Mater. 2016;32(2):285–93.
19. Hench LL. An introduction to bioceramics. London: Imperial College Press; 2013.
20. Skrtic D, Hailer AW, Takagi S, Antonucci JM, Eanes ED. Quantitative assessment of the efficacy of amorphous calcium phosphate/methacrylate composites in remineralizing caries-like lesions artificially produced in bovine enamel. J Dent Res. 1996;75(9):1679–86.
21. Langhorst SE, O'Donnell JN, Skrtic D. In vitro remineralization of enamel by polymeric amorphous calcium phosphate composite: quantitative microradiographic study. Dent Mater. 2009;25(7):884–91.
22. Weir MD, Chow LC, Xu HH. Remineralization of demineralized enamel via calcium phosphate nanocomposite. J Dent Res. 2012;91(10):979–84.
23. Melo MA, Weir MD, Rodrigues LK, Xu HH. Novel calcium phosphate nanocomposite with caries-inhibition in a human in situ model. Dent Mater. 2013;29(2):231–40.
24. Dickens SH, Flaim GM, Takagi S. Mechanical properties and biochemical activity of remineralizing resin-based Ca-PO4 cements. Dent Mater. 2003;19(6):558–66.
25. Dickens SH, Flaim GM. Effect of a bonding agent on in vitro biochemical activities of remineralizing resin-based calcium phosphate cements. Dent Mater. 2008;24(9):1273–80.
26. Peters MC, Bresciani E, Barata TJ, Fagundes TC, Navarro RL, Navarro MF, et al. In vivo dentin remineralization by calcium-phosphate cement. J Dent Res. 2010;89(3):286–91.

27. Weir MD, Ruan J, Zhang N, Chow LC, Zhang K, Chang X, et al. Effect of calcium phosphate nanocomposite on in vitro remineralization of human dentin lesions. Dent Mater. 2017;33(9):1033–44.
28. Moreau JL, Sun L, Chow LC, Xu HH. Mechanical and acid neutralizing properties and bacteria inhibition of amorphous calcium phosphate dental nanocomposite. J Biomed Mater Res B Appl Biomater. 2011;98(1):80–8.
29. Ionescu AC, Hahnel S, Cazzaniga G, Ottobelli M, Braga RR, Rodrigues MC, et al. Streptococcus mutans adherence and biofilm formation on experimental composites containing dicalcium phosphate dihydrate nanoparticles. J Mater Sci Mater Med. 2017;28(7):108.
30. Cheng L, Weir MD, Limkangwalmongkol P, Hack GD, Xu HH, Chen Q, et al. Tetracalcium phosphate composite containing quaternary ammonium dimethacrylate with antibacterial properties. J Biomed Mater Res B Appl Biomater. 2012;100(3):726–34.
31. Chen C, Weir MD, Cheng L, Lin NJ, Lin-Gibson S, Chow LC, et al. Antibacterial activity and ion release of bonding agent containing amorphous calcium phosphate nanoparticles. Dent Mater. 2014;30(8):891–901.
32. Melo MA, Cheng L, Zhang K, Weir MD, Rodrigues LK, Xu HH. Novel dental adhesives containing nanoparticles of silver and amorphous calcium phosphate. Dent Mater. 2013;29(2):199–210.
33. Cheng L, Weir MD, Xu HH, Antonucci JM, Lin NJ, Lin-Gibson S, et al. Effect of amorphous calcium phosphate and silver nanocomposites on dental plaque microcosm biofilms. J Biomed Mater Res B Appl Biomater. 2012;100(5):1378–86.
34. Cheng L, Weir MD, Xu HH, Kraigsley AM, Lin NJ, Lin-Gibson S, et al. Antibacterial and physical properties of calcium-phosphate and calcium-fluoride nanocomposites with chlorhexidine. Dent Mater. 2012;28(5):573–83.
35. Melo MA, Cheng L, Weir MD, Hsia RC, Rodrigues LK, Xu HH. Novel dental adhesive containing antibacterial agents and calcium phosphate nanoparticles. J Biomed Mater Res B Appl Biomater. 2013;101(4):620–9.
36. Natale LC, Alania Y, Rodrigues MC, Simoes A, de Souza DN, de Lima E, et al. Synthesis and characterization of silver phosphate/calcium phosphate mixed particles capable of silver nanoparticle formation by photoreduction. Mater Sci Eng C Mater Biol Appl. 2017;76:464–71.
37. Hench LL, Jones JR. Bioactive glasses: frontiers and challenges. Front Bioeng Biotechnol. 2015; 3:194.
38. Oguntebi B, Clark A, Wilson J. Pulp capping with bioglass and autologous demineralized dentin in miniature swine. J Dent Res. 1993;72(2):484–9.
39. Zamet JS, Darbar UR, Griffiths GS, Bulman JS, Bragger U, Burgin W, et al. Particulate bioglass as a grafting material in the treatment of periodontal intrabony defects. J Clin Periodontol. 1997;24(6):410–8.
40. Wefel JS. NovaMin: likely clinical success. Adv Dent Res. 2009;21(1):40–3.
41. Efflandt SE, Magne P, Douglas WH, Francis LF. Interaction between bioactive glasses and human dentin. Journal of materials science. Mater Med. 2002;13(6):557–65.
42. Greenspan DC, Hench LL. Bioactive glass for tooth remineralization. In: Hench LL, editor. An introduction to bioceramics. London: Impreial College Press; 2013.
43. Forsback AP, Areva S, Salonen JI. Mineralization of dentin induced by treatment with bioactive glass S53P4 in vitro. Acta Odontol Scand. 2004;62(1): 14–20.
44. Vollenweider M, Brunner TJ, Knecht S, Grass RN, Zehnder M, Imfeld T, et al. Remineralization of human dentin using ultrafine bioactive glass particles. Acta Biomater. 2007;3(6):936–43.
45. Sauro S, Osorio R, Watson TF, Toledano M. Influence of phosphoproteins' biomimetic analogs on remineralization of mineral-depleted resin-dentin interfaces created with ion-releasing resin-based systems. Dent Mater. 2015;31(7):759–77.
46. Wang Z, Jiang T, Sauro S, Wang Y, Thompson I, Watson TF, et al. Dentine remineralization induced by two bioactive glasses developed for air abrasion purposes. J Dent. 2011;39(11):746–56.
47. Profeta AC, Mannocci F, Foxton RM, Thompson I, Watson TF, Sauro S. Bioactive effects of a calcium/sodium phosphosilicate on the resin-dentine interface: a microtensile bond strength, scanning electron microscopy, and confocal microscopy study. Eur J Oral Sci. 2012;120(4):353–62.
48. Osorio R, Yamauti M, Sauro S, Watson TF, Toledano M. Experimental resin cements containing bioactive fillers reduce matrix metalloproteinase-mediated dentin collagen degradation. J Endod. 2012;38(9):1227–32.
49. Tezvergil-Mutluay A, Seseogullari-Dirihan R, Feitosa VP, Cama G, Brauer DS, Sauro S. Effects of composites containing bioactive glasses on demineralized dentin. J Dent Res. 2017;96(9):999–1005.
50. Chatzistavrou X, Velamakanni S, DiRenzo K, Lefkelidou A, Fenno JC, Kasuga T, et al. Designing dental composites with bioactive and bactericidal properties. Mater Sci Eng C Mater Biol Appl. 2015;52:267–72.
51. Yang SY, Kwon JS, Kim KN, Kim KM. Enamel surface with pit and fissure sealant containing 45S5 bioactive glass. J Dent Res. 2016;95(5):550–7.
52. Khvostenko D, Mitchell JC, Hilton TJ, Ferracane JL, Kruzic JJ. Mechanical performance of novel bioactive glass containing dental restorative composites. Dent Mater. 2013;29(11):1139–48.
53. Fernandes JS, Gentile P, Pires RA, Reis RL, Hatton PV. Multifunctional bioactive glass and glass-ceramic biomaterials with antibacterial properties for repair and regeneration of bone tissue. Acta Biomater. 2017;59:2–11.

54. Echezarreta-Lopez MM, Landin M. Using machine learning for improving knowledge on antibacterial effect of bioactive glass. Int J Pharm. 2013;453(2):641–7.
55. Stoor P, Soderling E, Salonen JI. Antibacterial effects of a bioactive glass paste on oral microorganisms. Acta Odontol Scand. 1998;56(3):161–5.
56. Allan I, Newman H, Wilson M. Antibacterial activity of particulate bioglass against supra- and subgingival bacteria. Biomaterials. 2001;22(12):1683–7.
57. Khvostenko D, Hilton TJ, Ferracane JL, Mitchell JC, Kruzic JJ. Bioactive glass fillers reduce bacterial penetration into marginal gaps for composite restorations. Dent Mater. 2016;32(1):73–81.
58. Torabinejad M, Hong CU, McDonald F, Pitt Ford TR. Physical and chemical properties of a new root-end filling material. J Endod. 1995;21(7):349–53.
59. Camilleri J. Hydration mechanisms of mineral trioxide aggregate. Int Endod J. 2007;40(6):462–70.
60. Gandolfi MG, Taddei P, Siboni F, Modena E, De Stefano ED, Prati C. Biomimetic remineralization of human dentin using promising innovative calcium-silicate hybrid "smart" materials. Dent Mater. 2011;27(11):1055–69.
61. Tay FR, Pashley DH, Rueggeberg FA, Loushine RJ, Weller RN. Calcium phosphate phase transformation produced by the interaction of the Portland cement component of white mineral trioxide aggregate with a phosphate-containing fluid. J Endod. 2007;33(11):1347–51.
62. Sarkar NK, Caicedo R, Ritwik P, Moiseyeva R, Kawashima I. Physicochemical basis of the biologic properties of mineral trioxide aggregate. J Endod. 2005;31(2):97–100.
63. Tay FR, Pashley DH. Guided tissue remineralisation of partially demineralised human dentine. Biomaterials. 2008;29(8):1127–37.
64. Tay FR, Pashley DH. Biomimetic remineralization of resin-bonded acid-etched dentin. J Dent Res. 2009;88(8):719–24.
65. Profeta AC, Mannocci F, Foxton R, Watson TF, Feitosa VP, De Carlo B, et al. Experimental etch-and-rinse adhesives doped with bioactive calcium silicate-based micro-fillers to generate therapeutic resin-dentin interfaces. Dent Mater. 2013;29(7):729–41.
66. Sauro S, Osorio R, Osorio E, Watson TF, Toledano M. Novel light-curable materials containing experimental bioactive micro-fillers remineralise mineral-depleted bonded-dentine interfaces. J Biomater Sci Polym Ed. 2013;24(8):940–56.
67. Osorio R, Yamauti M, Sauro S, Watson TF, Toledano M. Zinc incorporation improves biological activity of beta-tricalcium silicate resin-based cement. J Endod. 2014;40(11):1840–5.
68. Wang Z, Shen Y, Haapasalo M, Wang J, Jiang T, Wang Y, et al. Polycarboxylated microfillers incorporated into light-curable resin-based dental adhesives evoke remineralization at the mineral-depleted dentin. J Biomater Sci Polym Ed. 2014;25(7):679–97.
69. Profeta AC. Preparation and properties of calcium-silicate filled resins for dental restoration. Part II: micro-mechanical behaviour to primed mineral-depleted dentine. Acta Odontol Scand. 2014;72(8):607–17.
70. Torabinejad M, Hong CU, Pitt Ford TR, Kettering JD. Antibacterial effects of some root end filling materials. J Endod. 1995;21(8):403–6.
71. Damlar I, Ozcan E, Yula E, Yalcin M, Celik S. Antimicrobial effects of several calcium silicate-based root-end filling materials. Dent Mater J. 2014;33(4):453–7.
72. Parirokh M, Torabinejad M. Mineral trioxide aggregate: a comprehensive literature review—part I: chemical, physical, and antibacterial properties. J Endod. 2010;36(1):16–27.
73. Kim RJ, Kim MO, Lee KS, Lee DY, Shin JH. An in vitro evaluation of the antibacterial properties of three mineral trioxide aggregate (MTA) against five oral bacteria. Arch Oral Biol. 2015;60(10):1497–502.
74. Zhang H, Pappen FG, Haapasalo M. Dentin enhances the antibacterial effect of mineral trioxide aggregate and bioaggregate. J Endod. 2009;35(2):221–4.
75. Zehnder M, Waltimo T, Sener B, Soderling E. Dentin enhances the effectiveness of bioactive glass S53P4 against a strain of Enterococcus faecalis. Oral Surg Oral Med Oral Pathol Oral Radiol Endod. 2006;101(4):530–5.
76. Fan W, Wu Y, Ma T, Li Y, Fan B, et al. J Mater Sci Mater Med. 2016;27(1):16.
77. Fan W, Li Y, Sun Q, Ma T, Fan B. Calcium-silicate mesoporous nanoparticles loaded with chlorhexidine for both anti-Enterococcus faecalis and mineralization properties. J Nanobiotechnol. 2016;14(1):72.
78. Yang Y, Huang L, Dong Y, Zhang H, Zhou W, Ban J, et al. In vitro antibacterial activity of a novel resin-based pulp capping material containing the quaternary ammonium salt MAE-DB and Portland cement. PLoS One. 2014;9(11):e112549.
79. Cury JA, de Oliveira BH, dos Santos AP, Tenuta LM. Are fluoride releasing dental materials clinically effective on caries control? Dent Mater. 2016;32(3):323–33.
80. Hara AT, Turssi CP, Ando M, Gonzalez-Cabezas C, Zero DT, Rodrigues AL Jr, et al. Influence of fluoride-releasing restorative material on root dentine secondary caries in situ. Caries Res. 2006;40(5):435–9.
81. Cenci MS, Tenuta LM, Pereira-Cenci T, Del Bel Cury AA, ten Cate JM, Cury JA. Effect of microleakage and fluoride on enamel-dentine demineralization around restorations. Caries Res. 2008;42(5):369–79.
82. Ngo HC, Mount G, Mc Intyre J, Tuisuva J, Von Doussa RJ. Chemical exchange between glass-ionomer restorations and residual carious dentine in permanent molars: an in vivo study. J Dent. 2006;34(8):608–13.
83. Bezerra AC, Novaes RC, Faber J, Frencken JE, Leal SC. Ion concentration adjacent to glass-ionomer restorations in primary molars. Dent Mater. 2012;28(11):e259–63.

84. Kim YK, Yiu CK, Kim JR, Gu L, Kim SK, Weller RN, et al. Failure of a glass ionomer to remineralize apatite-depleted dentin. J Dent Res. 2010;89(3):230–5.
85. Duque C, Negrini Tde C, Hebling J, Spolidorio DM, et al. Oper Dent. 2005;30(5):636–40.
86. Davidovich E, Weiss E, Fuks AB, Beyth N. Surface antibacterial properties of glass ionomer cements used in atraumatic restorative treatment. J Am Dent Assoc (1939). 2007;138(10):1347–52.
87. van Dijken JW, Kalfas S, Litra V, Oliveby A. Fluoride and mutans streptococci levels in plaque on aged restorations of resin-modified glass ionomer cement, compomer and resin composite. Caries Res. 1997;31(5):379–83.
88. Bradshaw DJ, Marsh PD, Hodgson RJ, Visser JM. Effects of glucose and fluoride on competition and metabolism within in vitro dental bacterial communities and biofilms. Caries Res. 2002;36(2):81–6.
89. Vale GC, Tabchoury CP, Del Bel Cury AA, Tenuta LM, ten Cate JM, Cury JA. APF and dentifrice effect on root dentin demineralization and biofilm. J Dent Res. 2011;90(1):77–81.
90. Wiegand A, Buchalla W, Attin T. Review on fluoride-releasing restorative materials--fluoride release and uptake characteristics, antibacterial activity and influence on caries formation. Dent Mater. 2007;23(3):343–62.
91. Frencken JE, Imazato S, Toi C, Mulder J, Mickenautsch S, Takahashi Y, et al. Antibacterial effect of chlorhexidine-containing glass ionomer cement in vivo: a pilot study. Caries Res. 2007;41(2):102–7.
92. Yesilyurt C, Er K, Tasdemir T, Buruk K, Celik D. Antibacterial activity and physical properties of glass-ionomer cements containing antibiotics. Oper Dent. 2009;34(1):18–23.
93. Elsaka SE, Hamouda IM, Swain MV. Titanium dioxide nanoparticles addition to a conventional glass-ionomer restorative: influence on physical and antibacterial properties. J Dent. 2011;39(9):589–98.
94. Ibrahim MA, Neo J, Esguerra RJ, Fawzy AS. Characterization of antibacterial and adhesion properties of chitosan-modified glass ionomer cement. J Biomater Appl. 2015;30(4):409–19.
95. Paiva L, Fidalgo TKS, da Costa LP, Maia LC, Balan L, Anselme K, et al. Antibacterial properties and compressive strength of new one-step preparation silver nanoparticles in glass ionomer cements (NanoAg-GIC). J Dent. 2018;69:102–9.

10 Dental Resin-Based Materials with Antibacterial Properties: Contact Inhibition and Controlled Release

Satoshi Imazato and Haruaki Kitagawa

Abstract

The basic mechanical, physical, esthetic, and bonding properties of dental resin-based materials have greatly improved with technological developments, and the current materials used for restorative, prosthetic, and preventive treatments show excellent clinical performance. The target of their continued development is therefore shifting towards bioactive functionality to prevent primary/secondary diseases or promote tissue regeneration. Among the several bioactive properties proposed to further enhance dental materials, antibacterial effects that contribute to controlling bacterial infection are one of the most popular. Two approaches are available for conferring antibacterial activity to dental resin-based materials. One is immobilization of antimicrobial components in/on the materials that demonstrate the so-called contact inhibition—inhibition of bacteria that come into contact with the surfaces without any active components being released. Such technology involves utilization of a polymerizable bactericide such as quaternary ammonium compound (QAC)-based resin monomers or QAC-functionalized nanoparticles. In the other approach, the ability to liberate antibacterial components through controlled release is introduced using a carrier, e.g., nonbiodegradable polymer particles loaded with antimicrobials, silver nanoparticles, and ion-releasing glass fillers. In this chapter, two different approaches to providing dental resins with antibacterial properties are summarized. In addition, the future perspectives for each material are addressed based on the continued development of each approach. Both technologies for achieving antibacterial materials described here have great potential to contribute to successful dental treatments.

S. Imazato (✉) · H. Kitagawa
Department of Biomaterials Science,
Osaka University Graduate School of Dentistry,
Suita, Osaka, Japan
e-mail: imazato@dent.osaka-u.ac.jp;
h-kita@dent.osaka-u.ac.jp

10.1 Introduction

The need for new-generation dental materials with bioactive functions is a topic of great interest that is discussed not only by researchers but also by dental practitioners. Among the several bioactive properties proposed as beneficial enhancements for dental materials, antibacterial effects that contribute to controlling bacterial infection are some of the most popular.

Two major approaches are available for introducing antibacterial activity into dental resins. One is immobilization of antimicrobial components in/on the materials (Fig. 10.1a). This type of material demonstrates the so-called contact

A. C. Ionescu, S. Hahnel (eds.), *Oral Biofilms and Modern Dental Materials*,
https://doi.org/10.1007/978-3-030-67388-8_10

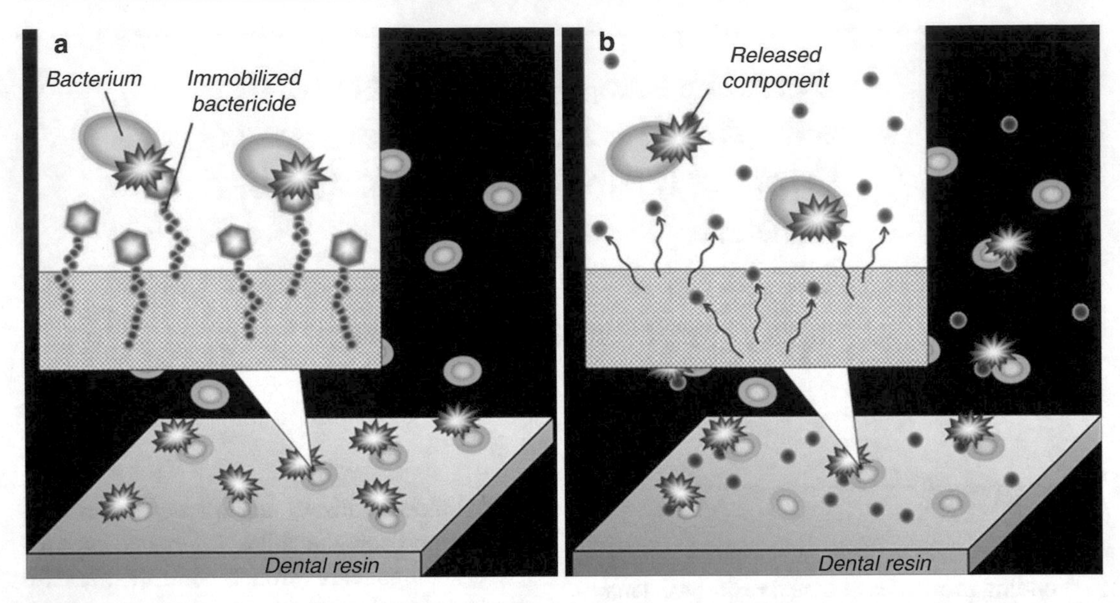

Fig. 10.1 Antimicrobial properties conferred by contact inhibition of immobilized bactericide (**a**) and controlled release of antimicrobial components (**b**)

inhibition, which means inhibition of bacteria that come into contact with its surface without any active components being released. The technology that has been intensively investigated to achieve immobilization of antibacterial components in dental resins is the use of polymerizable bactericides such as quaternary ammonium compound (QAC)-based resin monomers. Imazato et al. invented the world's first antibacterial QAC monomer, 12-methacryloyloxydodecylpyridinium bromide (MDPB) [1, 2], and many QAC monomers have subsequently been developed [3]. Dental resins with QAC-functionalized nanofillers [4, 5] have also been reported as another mode of QAC immobilization.

The other major approach is introducing into resins the ability to liberate antibacterial components by controlled release. The application of various filler particles as drug carriers has been investigated for this purpose [6] (Fig. 10.1b). It is not possible to achieve controlled-release kinetics using the classical method of mixing antimicrobial agents directly into materials, and the time span in which such materials exhibit antimicrobial effects is short. The use of drug carriers is an effective approach for overcoming such limitations and achieving long-lasting antimicrobial effects.

In this chapter, two different approaches for introducing antibacterial activity into dental resin-based materials are summarized along with the future perspectives for each material.

10.2 Contact Inhibition of Microorganisms by Immobilized Antimicrobials

10.2.1 Polymerizable Bactericide: QAC-Based Antibacterial Monomer MDPB

In 1994, Imazato et al. reported the first polymerizable bactericide for dental resins: antibacterial resin monomer MDPB. MDPB was synthesized by combining alkylpyridinium, a type of QAC, with a methacryloyl group [1] (Fig. 10.2). The QAC in the MDPB molecule is responsible for antibacterial activity, while the methacryloyl group allows for copolymerization with other conventional monomers. The antibacterial component is immobilized in the resin matrix after polymerization of MDPB (Fig. 10.3). The immobilized antimicrobial is not hydrolyzed and the antimicrobial components do not leach out in a wet environment. Therefore, cured resins con-

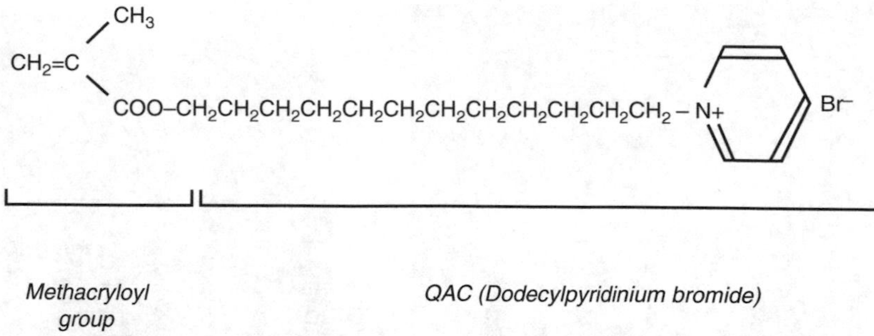

Fig. 10.2 Antibacterial resin monomer MDPB

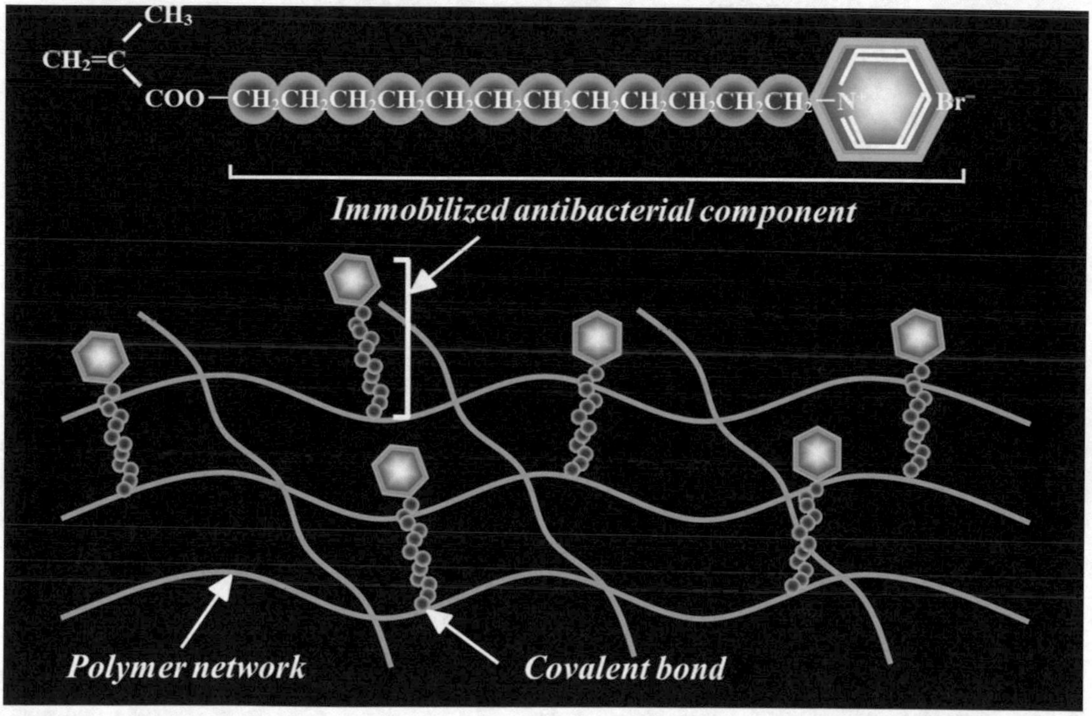

Fig. 10.3 Immobilized MDPB in a resin matrix

taining MDPB exhibit contact inhibition at their surfaces without detriment to their mechanical properties after function in the mouth.

10.2.2 Application of MDPB to Restorative Materials

Dental caries has been recognized as an infectious disease induced by cariogenic bacteria. Attempts to create restorative materials with antibacterial effects have provided an important topic in dental materials science. The control of bacteria around/beneath restorations could be advantageous for eliminating the risk of further demineralization and cavitation, contributing to the prevention of secondary caries. To address the growing need for antibacterial restorative materials, efforts have been made to apply the antibacterial monomer MDPB to several different materials.

The representative example of MDPB use is for adhesive systems. Experimental self-etching primer was prepared by incorporating MDPB, and its antibacterial activity and cavity-

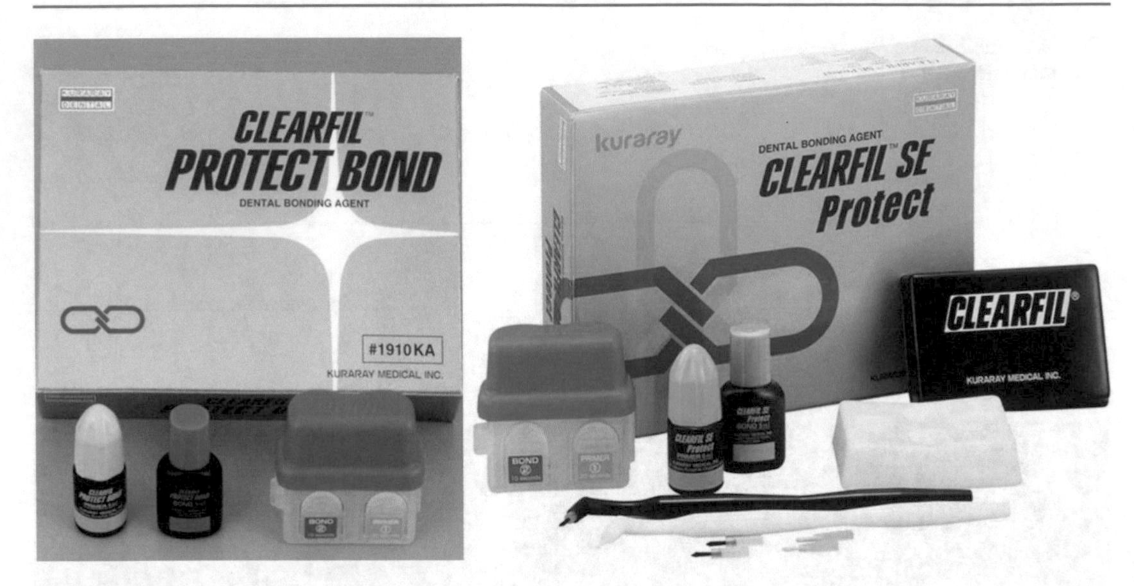

Fig. 10.4 The world's first antibacterial adhesive system Clearfil Protect Bond and its updated version Clearfil SE Protect

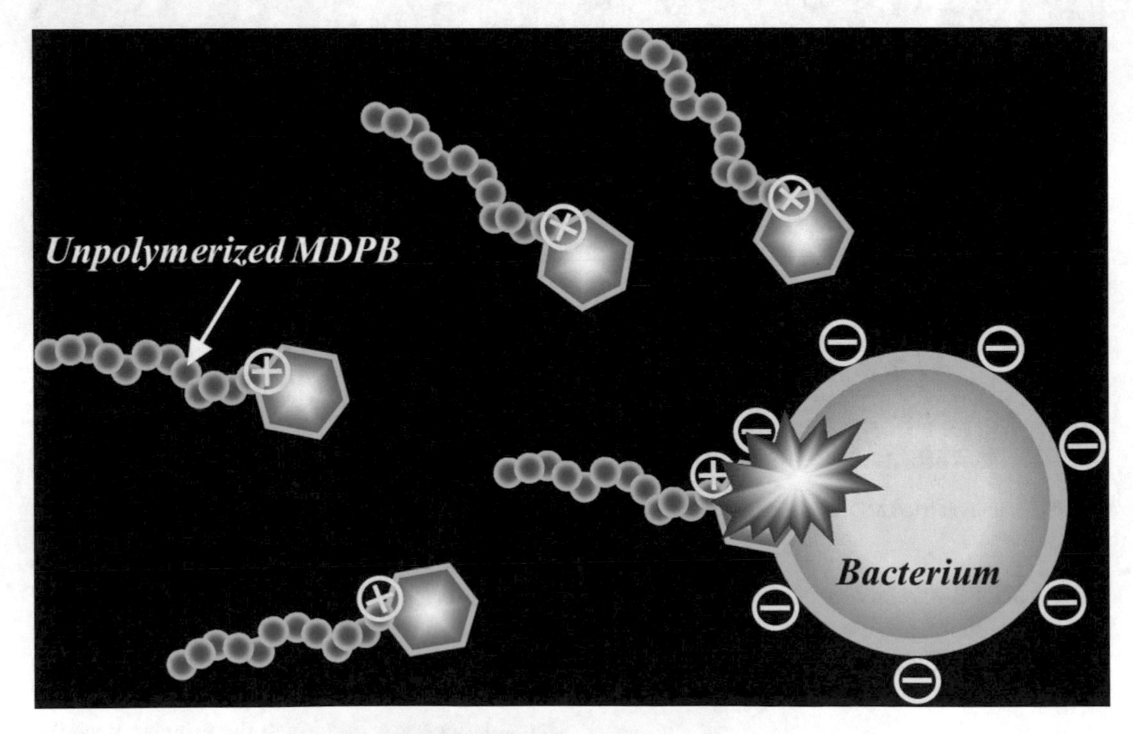

Fig. 10.5 The action of unpolymerized MDPB (a QAC) on a bacterium

disinfecting effects were confirmed by a number of in vitro and in vivo studies. Based on these findings, Clearfil Protect Bond (present product name: Clearfil SE Protect, Kuraray Noritake Dental Inc., Japan; Fig. 10.4), which employs a 5% MDPB-containing self-etching primer, was commercialized as the world's first antibacterial adhesive system.

MDPB-containing primer can kill bacteria rapidly based on the mechanism of action of unpolymerized MDPB—as a QAC—on bacteria (Fig. 10.5). When the primer containing MDPB is kept in direct contact with planktonic bacteria, effective bacterial killing can be attained within 30 s. It is noteworthy that the Clearfil Protect Bond primer can penetrate a 500-μm-thick dentin block and eradicate caries-related species inside the dentin [2]. In vivo studies using beagle dogs revealed that the MDPB-containing primer inactivated bacteria in the cavity [7]. Since residual bacteria are one of the primary causes of secondary caries, the cavity-disinfecting effects of the MDPB-containing primer are expected to improve the outcomes of restorative treatments of caries lesions. In addition, Brambilla et al. found from in vitro studies that biofilm formation around the margin of restorations placed using Clearfil Protect Bond was reduced owing to "contact inhibition" shown by immobilized MDPB [8].

The original aim of developing antibacterial monomer MDPB was for development of anti-plaque composites. MDPB was first immobilized in the resin matrix of composites [1], and to increase the concentration of immobilized bactericide, the prepolymerized resin filler containing MDPB was fabricated [9]. The composites incorporating MDPB-immobilized filler were reported to exhibit inhibitory effects against the growth of *Streptococcus mutans* on its surface and reduce plaque accumulation in vitro.

Recently, applications of MDPB have been expanded to other resinous materials such as a root canal sealer [10] and cavity disinfectant [11]. The experimental root canal sealer, which consisted of HEMA-based chemically cured primer containing MDPB at 5% and Bis-GMA-based sealing resin, was effective for killing the bacteria inside the dentinal tubules of root dentin and demonstrated excellent sealing ability. The experimental cavity disinfectant, intended for use for various direct and indirect restorations and prepared by adding 5% MDPB to 80% ethanol, was more effective in eradicating bacterial infection in dentin than the commercially available chlorhexidine-based cavity disinfectant. These findings confirmed that unpolymerized MDPB can penetrate deeply into dentinal tubules and exhibit strong bactericidal activity, which is expected to contribute to better clinical results in endodontic and restorative treatments.

10.2.3 Various QAC Monomers

Based on the concept of MDPB, new QAC monomers have subsequently been developed (Fig. 10.6). Huang et al. [12] synthesized 2-methacryloxylethyl dodecyl methyl ammonium bromide (MAE-DB) as a QAC with dimethacrylate groups. The cured experimental Bis-GMA/TEGDMA resin containing MAE-DB at 10 (wt)% demonstrated antibacterial effects against *Streptococcus mutans* even after immersion in water for 180 days. The low-viscosity ionic dimethacrylate monomer, bis(2-methacryloyloxy ethyl)dimethylammonium bromide (IDMA-1), was synthesized by the Menshutkin reaction [13]. Incorporation of 10 (wt)% IDMA-1 into Bis-GMA/TEGDMA resin reduced bacterial colonization without affecting the viability or metabolic activity of mammalian cells.

Li et al. [14] investigated the structure-property relationship of QAC monomers. A series of QAC monomers with alkyl chain lengths of 3, 6, 12, 16, and 18 were synthesized and their antimicrobial activities were evaluated. For short-chained quaternary ammonium compounds, the

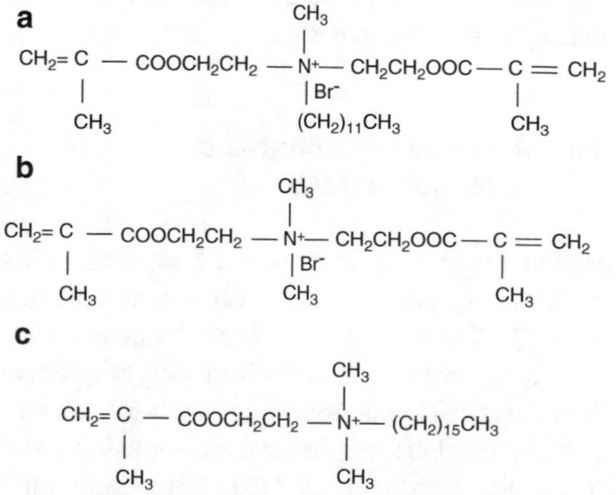

a

CH_3
$CH_2{=}C(CH_3)—COOCH_2CH_2—N^+(CH_3)((CH_2)_{11}CH_3)Br^-—CH_2CH_2OOC—C(CH_3){=}CH_2$

b

$CH_2{=}C(CH_3)—COOCH_2CH_2—N^+(CH_3)(CH_3)Br^-—CH_2CH_2OOC—C(CH_3){=}CH_2$

c

$CH_2{=}C(CH_3)—COOCH_2CH_2—N^+(CH_3)(CH_3)—(CH_2)_{15}CH_3$

Fig. 10.6 QAC monomers: (**a**) MAE-DB, (**b**) IDMA-1, (**c**) DMAHDM

antimicrobial activity appeared to rely on positively charged ammonium groups coupling with negatively charged bacterial membranes to disrupt membrane functions, alter the balance of essential ions, interrupt protein activity, and damage bacterial DNA. Increasing chain length reduced the metabolic activity and acid production of saliva-derived microcosm biofilms. Among the monomers with different alkyl chain lengths, the molecules with a chain length of 12 or 16, such as dimethylaminohexadecyl methacrylate (DMAHDM, Fig. 10.6c), were found to possess stronger antimicrobial effects.

Currently, many different QAC monomers are available including methacryloxylethyl cetyl dimethyl ammonium chloride (DMAE-CB), dimethylaminododecyl methacrylate (DMADDM), dimethylammoniumethyl dimethacrylate (DMAEDM), 2-dimethylamino ethyl methacrylate (DMAEMA), 2-methacryloxylethyl hexadecyl methyl ammonium bromide (MAE-HB), IDMA-2, methacryloyloxyundecylpyridinium bromide (MUPB), *N*-benzyl-11-(methacryloyloxy)-*N*,*N*-dimethylundecan-1-aminium fluoride (monomer II), [2-(methacryloyloxy)ethyl] trimethylammonium chloride (MADQUAT), urethane dimethacrylate quaternary ammonium methacrylate (UDMQA), and quaternary ammonium bisphenol A glycerolate dimethacrylate (QABGMA) [3, 6]. The effectiveness against oral microorganisms of the incorporation of each monomer, without impeding the physical properties, was demonstrated using in vitro assessments.

10.2.4 QAC-Functionalized Nanoparticles

Based on the development of various QAC monomers, antibacterial solid nanoparticles with QAC functionality have also been fabricated. Quaternary ammonium polyethyleneimine (QPEI) nanoparticles exhibit strong antibacterial effects against gram-positive and gram-negative bacteria [15], with their antimicrobial potency being attributed to the abundance of quaternary ammonium groups along the polymer backbone. Owing to their small size and large surface area, incorporation of a small amount of the QAC nanoparticles is sufficient to confer antibacterial effects and addition of 1–2 (wt)% provides contact killing effects to various resin-based materials [16–18]. Several recent studies have focused on surface modification of quaternary ammonium silica-based nanoparticles (QASi) to achieve QAC-functionalized nanoparticles [5]. These nanoparticles can be distributed more homogenously in polymers with less aggregation compared with QPEI nanoparticles.

10.2.5 Advantages and Disadvantages

Bactericide immobilized by polymerization of the antibacterial monomer is covalently bound to the base resins and does not leach out. Therefore, dental resins containing the immobilized bactericide demonstrate long-lasting inhibitory effects against microorganisms on their surfaces. MDPB-containing composites were confirmed to exhibit the same level of inhibition of biofilm formation when tested as cured and after 1 year of immersion in water.

As the immobilized component has limited molecular movement and disrupts the bacterial surface structure through contact action, its effect is essentially bacteriostatic rather than bactericidal. To obtain killing effects, the density of immobilized component exposed on the outer surface must be high, as shown by QPEI nanoparticles. A further disadvantage of the strategy of immobilizing bactericide in resins is that antibacterial effects are greatly reduced when the surface is covered by salivary protein. This is a critical problem for restorative materials used in the oral environment, and different key technologies need to be combined to achieve clinically effective anti-plaque activity.

10.3 Controlled Release of Antibacterial Components

10.3.1 Nonbiodegradable Polymer Particles Loaded with Antimicrobials

Simple approaches for adding antimicrobial components that get released in a wet environment to resinous materials have been reported for many years. However, the release behavior of the antimicrobial agents cannot be controlled using these methods and continuous delivery of the agents is challenging.

Imazato et al. [19] developed nonbiodegradable polymer particles made from the hydrophilic monomer 2-hydroxyethyl methacrylate (HEMA) and a cross-linking monomer trimethylolpropane trimethacrylate (TMPT), and reported their utility as antimicrobial reservoirs. By modifying the ratio of HEMA and TMPT, the hydrophilicity and polymer network density can be controlled [20]. Kitagawa et al. investigated loading of cetylpyridinium chloride (CPC) into these polymer particles using two different methods [20]. One method was to immerse the particles in a CPC aqueous solution to take up CPC. While it was found that the polymer particles consisting of 50% HEMA/50% TMPT were useful for loading and inhibition of bacteria, the duration of release of CPC from the particles loaded using the immersion method was short. Another method was to add CPC powder to the HEMA/TMPT monomer mixture and cure it to produce polymers (Fig. 10.7). With this method, a marked extension of the release period to over 120 days was achieved. The experimental resin-based endodontic sealer containing CPC-premixed particles was shown to exhibit antibacterial effects for a long period.

Moreover, the recharging of CPC into HEMA/TMPT particles by exposure to CPC solution was possible, and persistent antimicrobial effects with sustained release of CPC were achieved (Fig. 10.8). Such protocols are useful for maintaining clean surfaces for resinous materials such as denture bases.

Fig. 10.7 Nonbiodegradable CPC-loaded polymer particles made from HEMA and TMPT

10.3.2 Silver Nanoparticles

Silver is known to have antibacterial, antifungal, and antiviral activity. Silver ions strongly interact with thiol groups of vital enzymes and inactivate them, causing DNA to lose the ability to replicate and leading to cell death. Therefore, throughout the long history of dentistry, mixing silver into restorative materials has often been tried in attempts to provide antimicrobial effects. However, this method is not effective for providing continuous delivery of silver ions. In addition, silver incorporation results in blackening of the materials and is not suitable for esthetic restorations.

To overcome these problems, the addition of silver (Ag) nanoparticles to resinous materials has been investigated. Ag nanoparticles work as a kind of reservoir of Ag^+ that can be released in a controlled manner at a steady rate, allowing for long-term antibacterial effects. However, Ag nanoparticles are difficult to be dispersed in a resin matrix since nano-sized particles tend to aggregate and agglomerate. Therefore, to prevent aggregation, Ag nanoparticles are stabilized by various functional groups on their surface using coating agents or stabilizers such as polymers, polysaccharides, or citrates [6].

Another approach for avoiding the aggregation of Ag nanoparticles in resinous materials is a technique for fabricating polymers with evenly

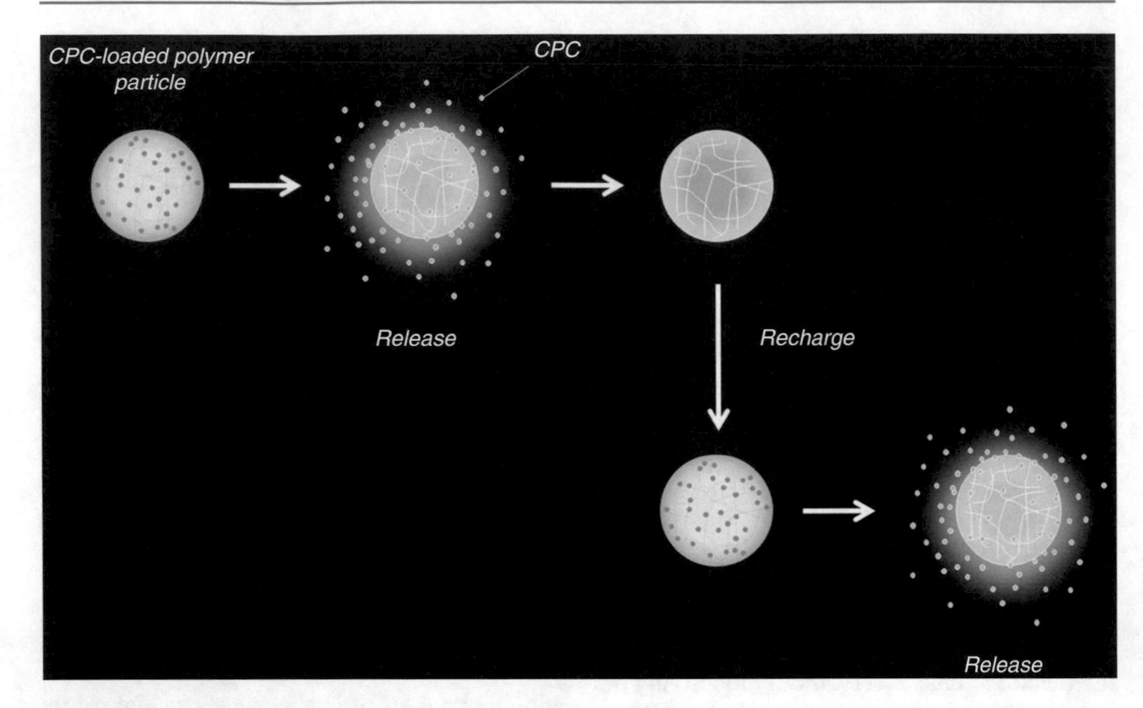

Fig. 10.8 Release and recharge profiles of CPC-loaded polymer particles

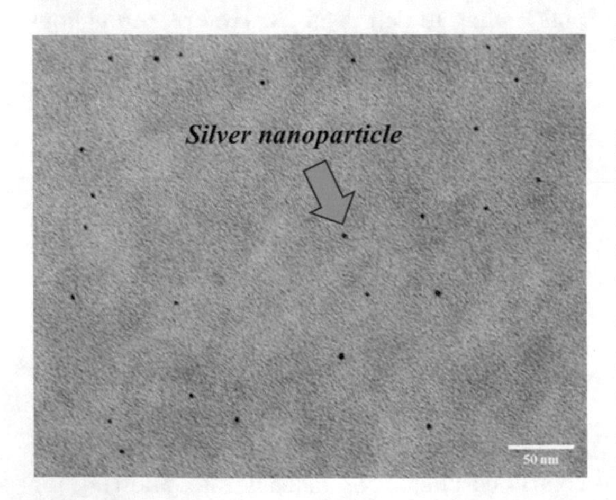

Fig. 10.9 TEM image of silver nanoparticles dispersed in the Bis-GMA/TEGDMA resin

dispersed Ag nanoparticles using coupling photoinitiated free radical polymerization of dimethacrylates with in situ silver ion reduction [21, 22]. Experimental resins prepared with this technology exhibit good Ag dispersion (Fig. 10.9), and strong reduction in bacteria coverage on their surface was observed in vitro for low-concentration addition of Ag nanoparticles [21].

10.3.3 Ion-Releasing Glass Fillers

10.3.3.1 Bioglass

Bioglass 45S5 (commercially known as Bioglass®), composed of SiO_2, Na_2O, CaO, and P_2O_5, is a potential candidate for use as antibacterial filler particles in restorative materials because it exerts antimicrobial effects as well as enhances hard tissue regeneration by releasing ions [6]. Its antimicrobial effects are attributed to the release of ions such as Ca^{2+}, which leads to a local increase in pH that is not well tolerated by bacteria. Resin composites containing Bioglass 65S, another type of bioglass composed of SiO_2, CaO, and P_2O_5, reduced bacterial penetration into the marginal gaps of simulated tooth restorations [23].

Davis et al. [24] developed glass fillers containing calcium and fluoride, prepared using the sol-gel method. It was shown that resin composites incorporating these fillers acted as a single source of both Ca^{2+} and F^- in aqueous solutions, and that the composites could be readily recharged with F^-. However, because the effec-

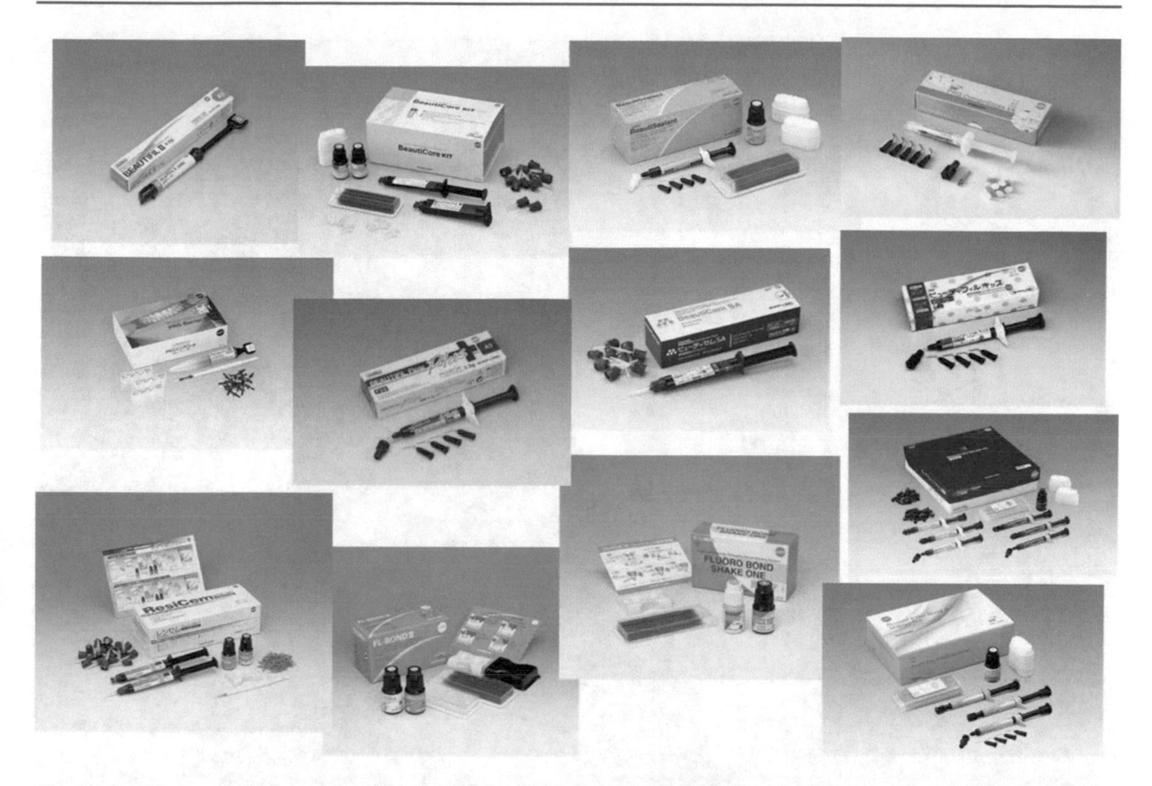

Fig. 10.10 Various resin-based materials containing S-PRG filler on the market

tive concentrations of Ca^{2+} and F^- against microorganisms are so high, the antimicrobial effects of a local increase in pH due to Ca^{2+} from bioglass, or the release of F^- from fluoride-containing glass fillers, are limited. As a result, additional components are needed to obtain apparent effects against oral microorganisms.

10.3.3.2 S-PRG Filler

Surface-pre-reacted glass-ionomer (S-PRG) filler is a material that releases multiple ions [2, 6], and many products containing S-PRG filler (SHOFU Inc., Japan) are already on the market (Fig. 10.10). This filler is prepared via an acid–base reaction between fluoro-boro-aluminosilicate glass and a polyacrylic acid. The pre-reacted glass-ionomer phase on the surface of the glass core allows S-PRG filler to release and recharge F^-. Moreover, S-PRG filler releases Al^{3+}, BO_3^{3-}, Na^+, SiO_3^{2-}, and Sr^{2+} ions (Fig. 10.11). Several studies have clearly demonstrated the antibacterial effects of resin composites containing S-PRG fillers against oral bacteria including *Streptococcus mutans*, and bacterial

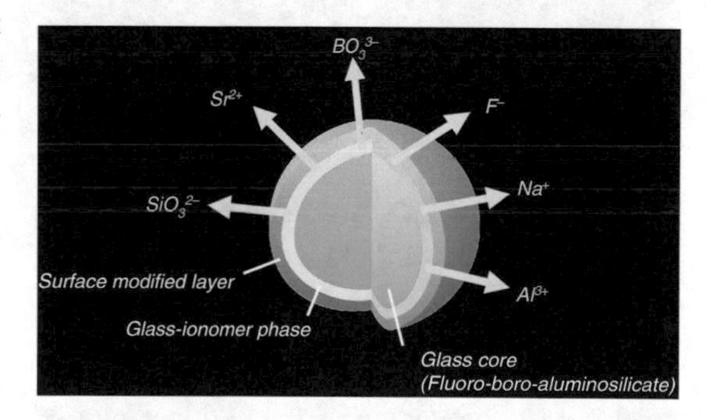

Fig. 10.11 The structure of S-PRG filler and the release of multiple ions

growth inhibition was obtained by release of BO_3^{3-} and F^- from the material. It was also reported that the eluate from the S-PRG filler suppressed the adherence of *Streptococcus mutans*, and thus biofilm formation on the surface of resin composites containing S-PRG fillers could be inhibited [6].

Recent investigations have revealed that *Streptococcus mutans* glucose metabolism and acid production were inhibited by low

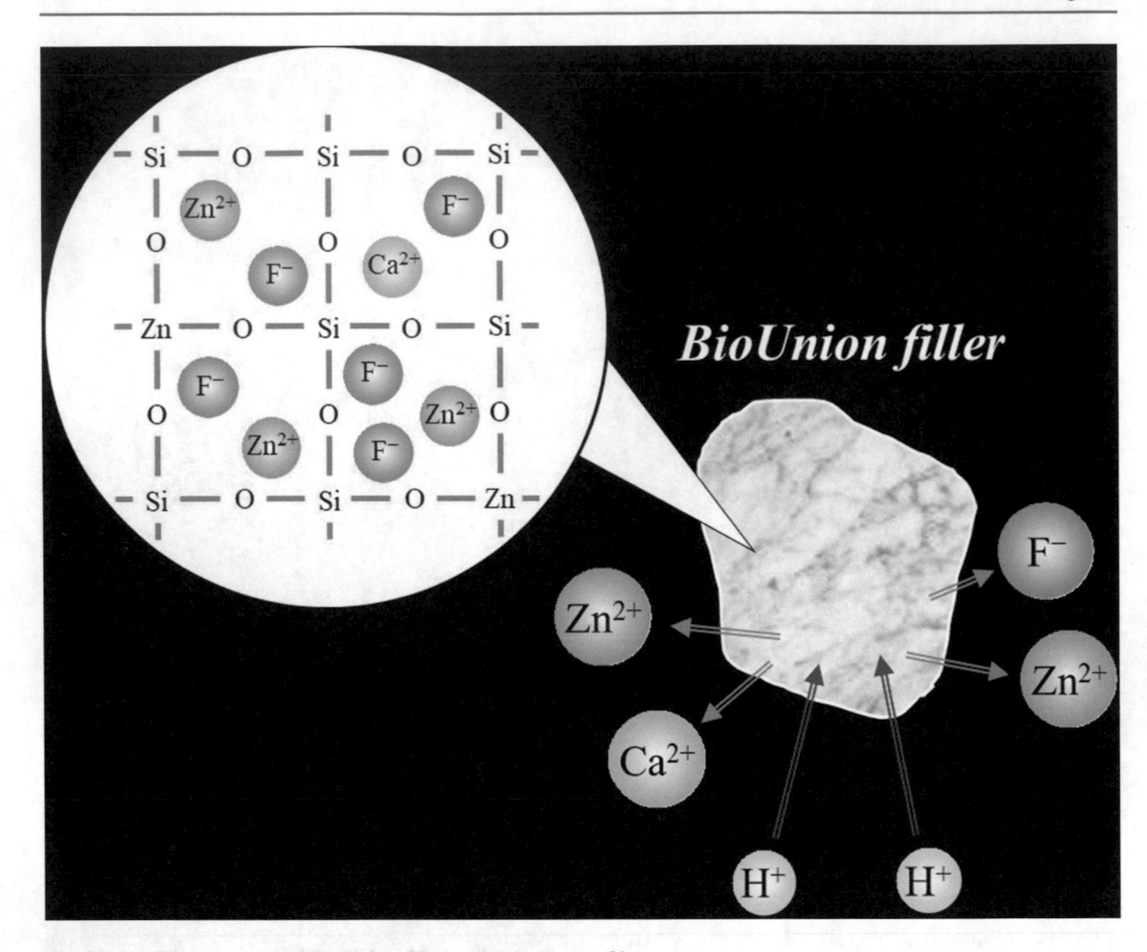

Fig. 10.12 The structure of BioUnion filler and the release of ions

concentrations of BO_3^{3-} or F^- that did not affect bacterial growth [25]. In addition, the eluate from S-PRG fillers effectively inhibited *Streptococcus mutans* growth through downregulation of genes related to sugar metabolism, resulting in attenuation of the cariogenicity of *Streptococcus mutans* [26]. Indeed, a coating resin containing S-PRG fillers, which produces a coating layer with a thickness of approximately 200 μm, inhibited the reduction in pH induced by glucose consumption of *Streptococcus mutans* on the material surface [27].

Ion release from S-PRG fillers is effective in suppressing the activity of periodontal pathogens. The eluate of S-PRG fillers shows inhibitory effects on the protease and gelatinase activity of *Porphyromonas gingivalis*. It has been demonstrated using animal studies that the eluate from S-PRG fillers exhibited preventive effects against tissue destruction in periodontal disease [28]. It was also found that the coaggregation of *P. gingivalis* and *Fusobacterium nucleatum* could be prevented in the presence of S-PRG filler eluate [29], indicating that the release of ions may contribute to the prevention of periodontitis.

10.3.3.3 BioUnion Filler

BioUnion filler is a glass powder composed of SiO_2, ZnO, CaO, and F, and can be categorized as a bio-functional multi-ion-releasing filler [6]. It has a silicon-based glass structure and is capable of releasing Zn^{2+}, Ca^{2+}, and F^- (Fig. 10.12). Zn^{2+} is known to exhibit antibacterial effects against oral bacteria, and its MIC/MBC values against *Streptococcus mutans* are lower than those of fluoride. Liu et al. [30] reported unique characteristics of BioUnion filler; that is, release of Zn^{2+} was accelerated under acidic conditions. Such technology enables on-demand release of antimicrobial components from the materials.

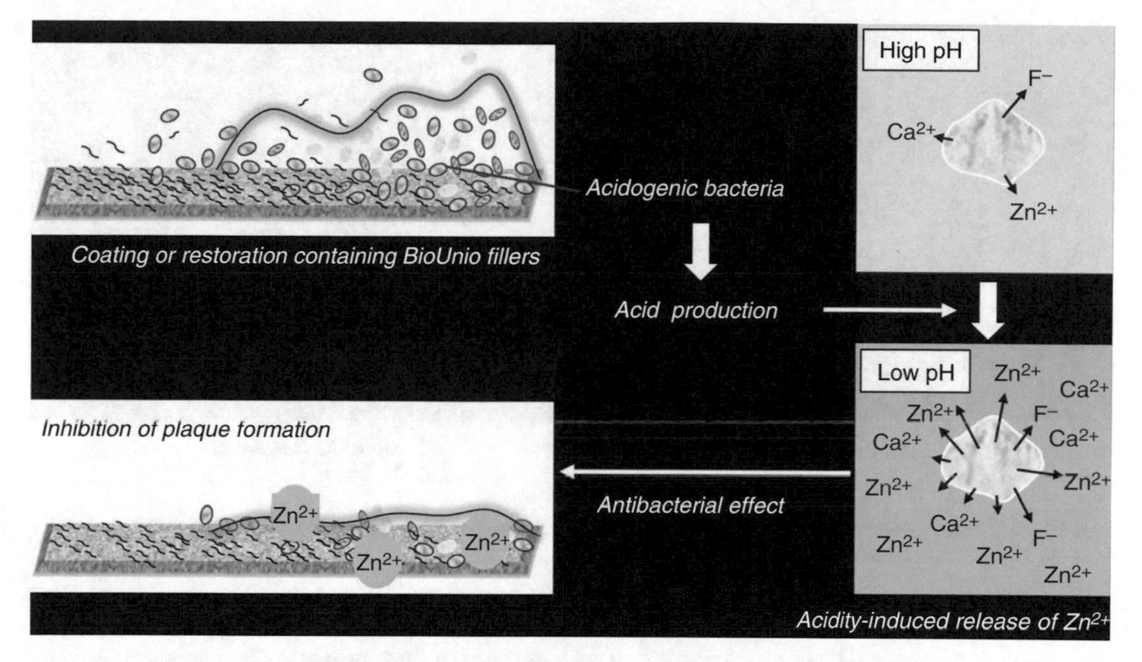

Fig. 10.13 Acidity-induced release of Zn^{2+} from BioUnion filler

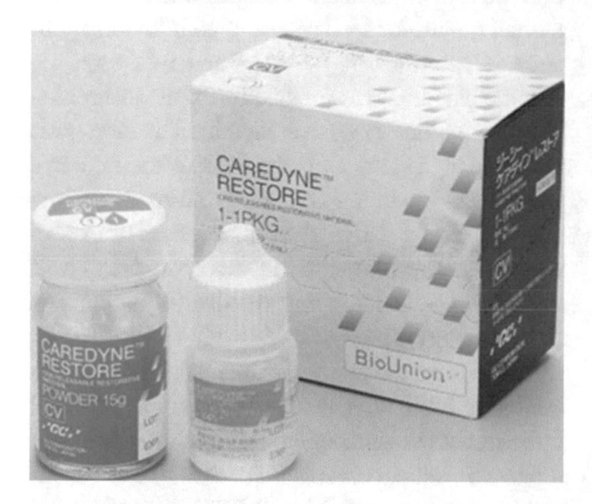

Fig. 10.14 Commercial product (dental cement) containing BioUnion filler

Once dental plaque is formed on the surface and acidogenic bacteria produce acids, a greater amount of Zn^{2+} is released and effectively attacks the cariogenic bacteria in the plaque (Fig. 10.13). While an inorganic cement for root surface restoration (Caredyne-Restore, GC Corporation, Japan) is on the market at present (Fig. 10.14), this technology is also of interest for use in resin-based materials.

10.3.4 Advantages and Disadvantages

Technologies for controlling release kinetics, such as the application of filler particles as drug reservoirs, are an effective way of enabling the sustained release of antibacterial components and achieving long-lasting antimicrobial effects. Particles capable of recharging the antibacterial component could be particularly practical for sustained release in clinical applications. In addition, the antibacterial effects of immobilized bactericide depend on direct contact with bacteria, while antimicrobial release systems are effective at inhibiting bacteria not only on the surface but also in some areas distant from the material. Therefore, the effectiveness in inhibiting biofilm formation for controlled antimicrobial release technology is essentially greater and reaches a much wider area.

The major disadvantage of the release of antimicrobials is the possibility of inducing microbial shift to disrupt homeostasis. Microbial shift in the oral cavity has been recognized as the help to cause infectious dental diseases. It is also important to assure biological safety regardless of the release of antimicrobials to attain clear, long-last-

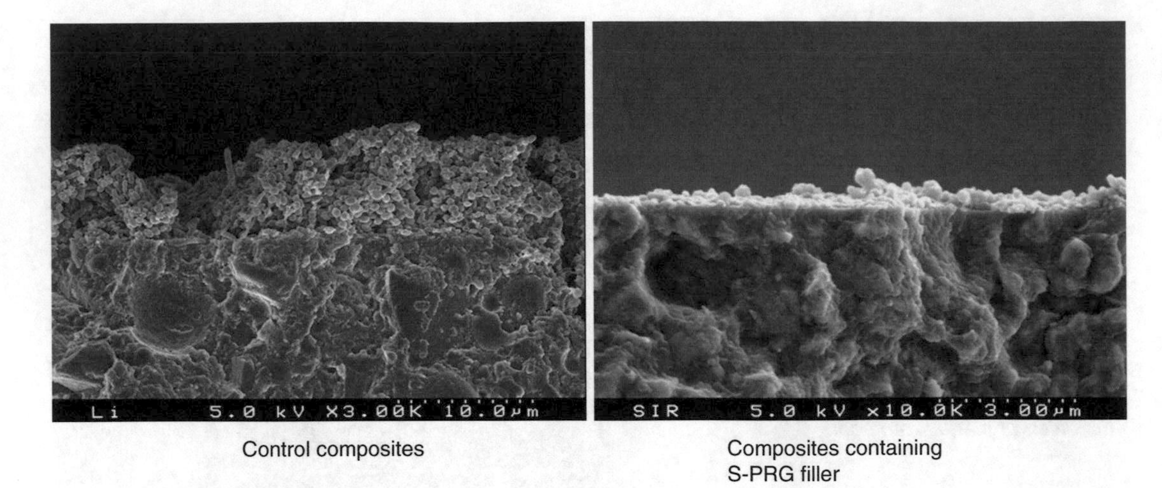

Fig. 10.15 Lower plaque accumulation on the surface of resin composites containing S-PRG filler after intraoral exposure for 24 h, compared with a control

ing antibacterial effects. This is not simply about toxicity to cells or organs, but also bacterial tolerance issues. It has been found that repeated exposure of *Enterococcus faecalis* and *Streptococcus gordonii* to chlorhexidine leads to resistance [31, 32], suggesting that sustained release of antimicrobials from materials may lead to drug resistance in oral bacteria.

10.4 Clinical Effectiveness

Despite the fact that many studies conducted in vitro or under clinically relevant experimental conditions have confirmed the benefits of both approaches—contact inhibition and controlled release—for providing antibacterial activity with dental resins, only a few clinical investigations are available so far.

For QAC-immobilized resins, the clinical effectiveness of MDPB-containing adhesive Clearfil Protect Bond (SE Protect) has been demonstrated by several studies. Uysal et al. [33] bonded orthodontic brackets to the premolars of 14 patients using Clearfil Protect Bond and examined demineralization of the enamel around the brackets after 30 days. They found that the hardness of the enamel around the brackets bonded with Clearfil Protect Bond was significantly greater than that around brackets bonded by other commercial orthodontic adhesives. The inhibitory effects of Clearfil Protect Bond against tooth demineralization in the case of composite restorations were also confirmed by in situ testing [34]. In this study, composite restoration was made in enamel blocks and kept in the human mouth for 14 days using an intraoral appliance. When Clearfil Protect Bond was used, no demineralization was found close to the restoration margin, while extensive demineralization was observed at the area away from the restoration. In a similar in situ study using a custom-made removable acrylic appliance, the effectiveness of experimental composites with incorporated QPEI nanoparticles in inhibiting biofilm formation was demonstrated [4].

Since there are many commercial products containing S-PRG filler on the market, their clinical effectiveness in terms of biofilm inhibition has been relatively well documented. For example, commercial composites containing S-PRG filler (Beautifil II; SHOFU Inc.) significantly inhibited plaque accumulation on their surface after intraoral exposure for 24 h compared with the control composites without S-PRG filler (Fig. 10.15) [35].

Although information obtained by in situ assessments reflects the significant effectiveness of antibacterial materials in clinical settings, such studies are labor and time consuming. In addition, it is not easy to collect a large amount of results. Ethical issues are also raised, particularly for testing experimental materials that release antimicrobial components. Therefore, an important future direction is to develop convenient in vitro evaluation systems that are specifically designed for antibacterial materials, with realistic simulations of the oral environment. More information on this topic is available in Chap. 4.

10.5 Future Perspectives

10.5.1 Further Improvement of the Immobilization Approach

10.5.1.1 Protein-Repellent Properties

Antibacterial component immobilized by polymerization of QAC monomers shows inhibition of bacteria and its effect can last for long periods as the immobilized component does not leach out. However, immobilized component exhibits antibacterial effects by contact inhibition; hence its effectiveness is reduced as a result of salivary protein coverage in the oral environment. One of the approaches for solving this problem is to introduce protein-repellent properties at the surface with immobilized bactericide. The application of molecules such as 2-methacryloyloxyethyl phosphorylcholine (MPC) polymer is promising. MPC polymer is hydrophilic and has an abundance of free water, but no bound water, when in the hydrated state. The large amount of free water around the phosphorylcholine groups is thought to detach proteins effectively, thereby repelling protein adsorption. The combination of MPC with QAC monomers has been reported to assist in impeding biofilm formation on dental materials [36, 37].

10.5.1.2 Grafting Approach

Resinous materials containing QAC monomer are able to exhibit bactericidal effects only when the density of immobilized QAC is high enough to disrupt bacterial cell structure. Grafting QAC monomers onto the surface has been shown to be effective for achieving restorative materials with strong effects [38]. Several techniques for grafting monomers on polymer surfaces are used in industry for surface modification and functionalization with polymers. However, most of these approaches require the addition of reactive functional groups or initiators to the grafted surfaces, and such complications make it difficult to apply the methods to dental materials for direct restoration. Simple methods for grafting QAC monomers on any type of resinous dental materials are therefore desirable.

10.5.2 Further Improvement of Controlled Release

Dental caries is a disease caused by acids from glucose metabolism of specific bacteria in the oral cavity. One of the aims of conferring antibacterial activities to restorative or preventive materials is to suppress caries-related bacteria and inhibit the pH decrease that occurs in dental plaque. The BioUnion fillers described above are capable of releasing Zn^{2+} when the environmental pH decreases. Similar responsive technology that produces antibacterial effects as a result of environmental (e.g., pH or microbiota) changes is of great interest. The future design of dental materials with agent release needs to involve "smart" behavior for the maintenance of the oral environment and to safeguard human health.

Acknowledgment We thank Dr. Hockin H.K. Xu for providing Fig. 10.8.

References

1. Imazato S, Torii M, Tsuchitani Y, et al. Incorporation of bacterial inhibitor into resin composite. J Dent Res. 1994;73:1437–43.
2. Imazato S, Ma S, Chen JH, et al. Therapeutic polymers for dental adhesives: loading resins with bioactive components. Dent Mater. 2014;30:97–104.
3. Jiao Y, Niu L-N, Ma S, et al. Quaternary ammonium-based biomedical materials: state-of-the-art, toxicological aspects and antimicrobial resistance. Prog Polym Sci. 2017;71:53–90.
4. Beyth N, Yudovin-Farber I, Perez-Davidi M, et al. Polyethyleneimine nanoparticles incorporated into resin composite cause cell death and trigger biofilm stress *in vivo*. Proc Natl Acad Sci. 2010;107:22038–43.
5. Zaltsman N, Ionescu AC, Weiss EI, et al. Surface-modified nanoparticles as anti-biofilm filler for dental polymers. PLoS One. 2017;12:1–18.
6. Imazato S, Kohno T, Tsuboi R, et al. Cutting-edge filler technologies to release bio-active components for restorative and preventive dentistry. Dent Mater J. 2020;39:69–79.
7. Imazato S, Kaneko T, Takahashi Y, et al. *In vivo* antibacterial effects of dentin primer incorporating MDPB. Oper Dent. 2004;29:369–75.

8. Brambilla E, Ionescu A, Fadini L, et al. Influence of MDPB-containing primer on *Streptococcus mutans* biofilm formation in simulated class I restorations. J Adhes Dent. 2013;15:431–8.
9. Imazato S, Ebi N, Takahashi Y, et al. Antibacterial activity of bactericide-immobilized filler for resin-based restoratives. Biomaterials. 2003;24:3605–9.
10. Kitagawa R, Kitagawa H, Izutani N, et al. Development of an antibacterial root canal filling system containing MDPB. J Dent Res. 2014;93:1277–82.
11. Hirose N, Kitagawa R, Kitagawa H, et al. Development of a cavity disinfectant containing antibacterial monomer MDPB. J Dent Res. 2016;95:1487–93.
12. Huang L, Sun X, Xiao YH, et al. Antibacterial effect of a resin incorporating a novel polymerizable quaternary ammonium salt MAE-DB against *Streptococcus mutans*. J Biomed Mater Res B Appl Biomater. 2012;100:1353–8.
13. Antonucci JM, Zeiger DN, Tang K, et al. Synthesis and characterization of dimethacrylates containing quaternary ammonium functionalities for dental applications. Dent Mater. 2012;28:219–28.
14. Li F, Weir MD, Xu HH. Effects of quaternary ammonium chain length on antibacterial bonding agents. J Dent Res. 2013;92:932–8.
15. Farah S, Aviv O, Laout N, et al. Quaternary ammoniumpolyethylenimine nanoparticles for treating bacterial contaminated water. Colloids Surf B. 2015;128:614–9.
16. Yudovin-Farber I, Beyth N, Nyska A, et al. Surface characterization and biocompatibility of restorative resin containing nanoparticles. Biomacromolecules. 2008;9:3044–50.
17. Shvero DK, Davidi MP, Weiss EI, et al. Antibacterial effect of polyethyleneimine nanoparticles incorporated in provisional cements against *Streptococcus mutans*. J Biomed Mater Res B Appl Biomater. 2010;94:367–71.
18. Shvero DK, Abramovitz I, Zaltsman N, et al. Towards antibacterial endodontic sealers using quaternary ammonium nanoparticles. Int Endod J. 2013;46:747–54.
19. Imazato S, Kitagawa H, Tsuboi R, et al. Non-biodegradable polymer particles for drug delivery: a new technology for "bio-active" restorative materials. Dent Mater J. 2017;36:524–32.
20. Kitagawa H, Takeda K, Kitagawa R, et al. Development of sustained antimicrobial-release systems using poly(2-hydroxyethyl methacrylate)/trimethylolpropane trimethacrylate hydrogels. Acta Biomater. 2014;10:4285–95.
21. Cheng Y-J, Zeiger DN, Howarter JA, et al. *In situ* formation of silver nanoparticles in photocrosslinking polymers. J Biomed Mater Res B Appl Biomater. 2011;97B:124–31.
22. Zhang K, Li F, Imazato S, et al. Dual antibacterial agents of nano-silver and 12-methacryloyloxy dodecylpyridinium bromide in dental adhesive to inhibit caries. J Biomed Mater Res B Appl Biomater. 2013;101B:929–38.
23. Khvostenko D, Hilton TJ, Ferracane JL, et al. Bioactive glass fillers reduce bacterial penetration into marginal gaps for composite restorations. Dent Mater. 2016;32:73–81.
24. Davis HB, Gwinner F, Mitchell JC, et al. Ion release from, and fluoride recharge of a composite with a fluoride-containing bioactive glass. Dent Mater. 2014;30:1187–94.
25. Kitagawa H, Miki-Oka S, Mayanagi G, et al. Inhibitory effect of resin composite containing S-PRG filler on *Streptococcus mutans* glucose metabolism. J Dent. 2018;70:92–6.
26. Nomura R, Morita Y, Matayoshi S, et al. Inhibitory effect of surface pre-reacted glass-ionomer (S-PRG) eluate against adhesion and colonization by *Streptococcus mutans*. Sci Rep. 2018;8:5056.
27. Ma S, Imazato S, Chen J-H, et al. Effects of a coating resin containing S-PRG filler to prevent demineralization of root surfaces. Dent Mater J. 2012;31:909–15.
28. Iwamatsu-Kobayashi Y, Abe S, Fujieda Y, et al. Metal ions from S-PRG filler have the potential to prevent periodontal disease. Clin Exp Dent Res. 2017;3:126–33.
29. Yoneda M, Suzuki N, Masuo Y, et al. Effect of S-PRG eluate on biofilm formation and enzyme activity of oral bacteria. Int J Dent. 2012;2012:1–6.
30. Liu Y, Kohno T, Tsuboi R, et al. Acidity-induced release of zinc ion from BioUnion filler™ and its inhibitory effects against *Streptococcus mutans*. Dent Mater J. 2020;39:547–53.
31. Kitagawa H, Izutani N, Kitagawa R, et al. Evolution of resistance to cationic biocides in *Streptococcus mutans* and *Enterococcus faecalis*. J Dent. 2016;47:18–22.
32. Wang S, Wang H, Ren B, et al. Drug resistance of oral bacteria to new antibacterial dental monomer dimethylaminohexadecyl methacrylate. Sci Rep. 2018;8:5509.
33. Uysal T, Amasyali M, Ozcan S, et al. Effect of antibacterial monomer-containing adhesive on enamel demineralization around orthodontic brackets: an *in-vivo* study. Am J Orthod Dentofac Orthop. 2011;139:650–6.
34. Pinto CF, Berger SB, Cavalli V, et al. *In situ* antimicrobial activity and inhibition of secondary caries of self-etching adhesives containing an antibacterial agent and/or fluoride. Am J Dent. 2015;28:167–73.
35. Saku S, Kotake H, Scougall-vilchis RJ, et al. Antibacterial activity of composite resin with glass-ionomer filler particles. Dent Mater J. 2010;29:193–8.
36. Zhang K, Zhang K, Xie X, et al. Nanostructured polymeric materials with protein-repellent and anti-caries properties for dental applications. Nanomaterials. 2018;8:E393.
37. Thongthai P, Kitagawa H, Kitagawa R, et al. Development of novel surface coating composed of MDPB and MPC with dual functionality of antibacterial activity and protein repellency. J Biomed Mater Res B Appl Biomater. 2020;108:3241–9.
38. Müller R, Eidt A, Hiller KA, et al. Influences of protein films on antibacterial or bacteria-repellent surface coatings in a model system using silicon wafers. Biomaterials. 2009;30:4921–9.

Conclusion

Like in many other aspects of modern materials science, the analysis of interactions between biofilms and dental materials is a quickly evolving and increasingly important scientific field. An effect of this circumstance is that the available knowledge is rapidly accumulating, while the interlapse between paradigm shifts is accordingly reducing. Biofilms have been colonising Earth for hundreds of millions of years and we, as species, have managed to coexist with them as much as the other way around. Host and its residents together contribute to health and disease as a holobiont—a group of different species forming an ecological unit [Lynn Margulis]. As a result, dental materials science has to broaden its perspective and take into account not only what is related to the mechanical performance and aesthetic characteristics of a restorative material but also its behavior in the modified ecological unit. Despite the latest advancements displayed in this book, we are still far from being able to provide perfect dental materials. However, we are more and more capable of understanding the interactions of materials with the oral environment, which results in the development of materials featuring at least some of the characteristics that are considered as necessary to ensure longevity in such extreme environments. New generations of dental materials will be able to resist biodeteriorating processes more effectively than older formulations, and some materials will also be capable to modulate the adherent salivary pellicle, biofilms, and surrounding natural tissues in a favourable manner. The European Society of Biomaterials defines biomaterials as a "material intended to interface with biological systems". Taking this definition into account, it may not be inappropriate to label future dental materials as dental biomaterials, which further underlines the complexity and variety of these modern materials.

A. C. Ionescu, S. Hahnel (eds.), *Oral Biofilms and Modern Dental Materials*,
https://doi.org/10.1007/978-3-030-67388-8

Printed in the United States
by Baker & Taylor Publisher Services